POWER SYSTEM HARMONICS - ANALYSIS, EFFECTS AND MITIGATION SOLUTIONS FOR POWER QUALITY IMPROVEMENT

Edited by **Ahmed Zobaa, Shady H. E. Abdel Aleem** and **Murat Erhan Balci**

Power System Harmonics - Analysis, Effects and Mitigation Solutions for Power Quality Improvement
http://dx.doi.org/10.5772/intechopen.68674
Edited by Ahmed Zobaa, Shady H. E. Abdel Aleem and Murat Erhan Balci

Contributors

Mohammed Albadi, Rashid Al-Abri, Amer Al-Hinai, Abdullah Al-Badi, Yang Du, Dylan Dah-Chuan Lu, Luís Monteiro, Alexey Tyshko, Ahmed F. F Zobaa, Shady H.E. Abdel Aleem, Murat Erhan Balci

Notice

Statements and opinions expressed in the chapters are these of the individual contributors and not necessarily those of the editors or publisher. No responsibility is accepted for the accuracy of information contained in the published chapters. The publisher assumes no responsibility for any damage or injury to persons or property arising out of the use of any materials, instructions, methods or ideas contained in the book.

First published in London, United Kingdom, 2018 by IntechOpen
IntechOpen is the global imprint of INTECHOPEN LIMITED, registered in England and Wales, registration number: 11086078, The Shard, 25th floor, 32 London Bridge Street
London, SE19SG – United Kingdom
Printed in Croatia

British Library Cataloguing-in-Publication Data
A catalogue record for this book is available from the British Library

Additional hard copies can be obtained from orders@intechopen.com

Power System Harmonics - Analysis, Effects and Mitigation Solutions for Power Quality Improvement, Edited by Ahmed Zobaa, Shady H. E. Abdel Aleem and Murat Erhan Balci
p. cm.
Print ISBN 978-1-78923-190-8
Online ISBN 978-1-78923-191-5

We are IntechOpen,
the world's leading publisher of
Open Access books
Built by scientists, for scientists

3,450+
Open access books available

110,000+
International authors and editors

115M+
Downloads

151
Countries delivered to

Our authors are among the

Top 1%
most cited scientists

12.2%
Contributors from top 500 universities

CLARIVATE ANALYTICS
BOOK
CITATION
INDEX
INDEXED

WEB OF SCIENCE™

Selection of our books indexed in the Book Citation Index
in Web of Science™ Core Collection (BKCI)

Interested in publishing with us?
Contact book.department@intechopen.com

Meet the editors

Ahmed Zobaa received his BSc (Hons), MSc, and PhD degrees in Electrical Power and Machines from Cairo University, Egypt, in 1992, 1997, and 2002, respectively. Also, he received his Postgraduate Certificate in Academic Practice from the University of Exeter, UK, in 2010. Also, he received his Doctor of Science degree from Brunel University London, UK, in 2017. He was an instructor during 1992–1997, a teaching assistant during 1997–2002, and an assistant professor during 2002–2007 at Cairo University, Egypt. From 2007 to 2010, he was a senior lecturer in renewable energy at the University of Exeter, UK. Currently, he is a senior lecturer in electrical and power engineering, an MSc course director, and a full member of the Institute of Energy Futures at Brunel University London, UK. His main areas of expertise include power quality, (marine) renewable energy, smart grids, energy efficiency, and lighting applications. Dr. Zobaa is an executive editor for the *International Journal of Renewable Energy Technology*. Also, he is an editor in chief for *Technology and Economics of Smart Grids and Sustainable Energy* and *International Journal of Electrical Engineering Education*. He is a senior fellow of the Higher Education Academy of UK. He is a fellow of the Institution of Engineering and Technology, the Energy Institute of UK, the Chartered Institution of Building Services Engineers, the Institution of Mechanical Engineers, the Royal Society of Arts, the African Academy of Science, and the Chartered Institute of Educational Assessors.

Shady H. E. Abdel Aleem received his PhD degree in Electrical Power and Machines from the Faculty of Engineering, Cairo University, Egypt, in 2013. Currently, he is an assistant professor at 15th of May Higher Institute of Engineering. His research interests include harmonic distortion problems in power systems, power quality, renewable energy, smart grid, and optimization. He has published over 60 journals and conference papers, 6 book chapters, and 3 edited books. Dr. Shady is a member of the Institute of Electrical and Electronics Engineers (IEEE) and Institution of Engineering and Technology (IET). He is an editor for the *International Journal of Renewable Energy Technology*, *Vehicle Dynamics*, and *International Journal of Electrical Engineering Education*.

Murat Erhan Balci received his BSc degree from Kocaeli University and MSc and PhD degrees from Gebze Institute of Technology, Turkey, in 2001, 2004, and 2009, respectively. During 2008, he was a visiting scholar at Worcester Polytechnic Institute, USA. Since 2009, he has been with the Electrical and Electronics Engineering Department of Balikesir University, Turkey. He is, currently, an associate professor at the same University. He is working in the field of electrical machines, power electronics, power quality, power system analysis, and renewable energy.

Contents

Preface

Power quality is related to magnitude, frequency, and waveform of voltage and current. For good power quality level, supply voltages and line currents should be within their rated magnitudes, the frequency close to the prescribed supply frequency and sinusoidal waveform. There are many events disturbing power quality performance such as over-undervoltages, harmonic distortion, flickers, unbalance, sags, swells, transients, interruptions and frequency deviations. In addition, harmonic distortion is recognized as one of the most important power quality events in the literature since it has many adverse effects on operation and control of power system equipment.

Harmonic distortion occurs when voltage and currents deviate from the sinusoidal waveform. For the phasor analysis of these distorted waveforms, using Fourier transform, they are separated to the sinusoidal components, which have the frequencies of multiple integers of the supply frequency, called harmonics. In modern power systems, the main sources of current harmonic distortion are the loads and renewable energy generation units, which are all connected into a system via power electronic interfaces. At the same time, voltage drops on the line impedances caused by the distorted currents lead to harmonic distortion of the bus voltages in the system.

The most important impacts of harmonics on the power system equipment are the overheating and torque oscillations of the induction motors, overheating and decreased power transfer capability of the transformers and supply lines and malfunctions of the protection/measurement devices. Thus, today, several international standards, such as IEEE Standard 519 and IEC 61000, present harmonic limitations for power systems. Accordingly, the harmonic mitigation has gained importance and passive, active or hybrid filters are widely employed in power systems to mitigate the adverse harmonic distortion effects.

This book aims to present harmonic modeling, analysis and mitigation techniques for modern power systems. It is a tool for the planners, designers, operators and practicing engineers of electrical power systems involved in the power system harmonics. Likewise, it is a key resource for advanced students, postgraduates, academics and researchers who have some background in electrical power systems. The book is sorted out and organized in five chapters.

Chapter 1: This chapter summarizes the power quality events and their mitigation techniques.

Chapter 2: This chapter provides a discussion involving new trends on distribution power grids with active power filters to improve power quality, increase the reliability of the power grid and contribute to make feasible the implementation of decentralized microgrids.

Chapter 3: This chapter focuses on the sequential harmonic elimination method employed for multimodule multilevel converters. The principles of the sequential selective harmonic elimination for MMC topology and amplitude control are described with examples.

Chapter 4: This chapter introduces a general model modified from the conventional control structure diagram for analysis of the harmonic generation process for photovoltaic (PV) installations. Causes of the current harmonics and their relationships with output power levels are summarized and analyzed.

Chapter 5: This chapter presents a case study about harmonic measurements in high-voltage networks. The measurements are analyzed and temporal harmonic profiles are studied in detail.

Ahmed Zobaa
College of Engineering
Design and Physical Sciences
Brunel University London
Uxbridge, United Kingdom

Shady H. E. Abdel Aleem
15th of May Higher Institute of Engineering
Mathematical and Physical Sciences
Cairo, Egypt

Murat Erhan Balci
Electrical and Electronics Engineering
Balikesir University
Balikesir, Turkey

Introductory Chapter: Power System Harmonics— Analysis, Effects, and Mitigation Solutions for Power Quality Improvement

Ahmed F. Zobaa, Shady H.E. Abdel Aleem and
Murat E. Balci

Additional information is available at the end of the chapter

http://dx.doi.org/10.5772/intechopen.76628

1. Introduction

Nowadays, electrical utilities and consumers are paying much attention to enhance the quality of the generated and distributed electrical energy. The main aims are to produce clean electrical power and to distribute it to the end customers with acceptable power quality performance in a cost-effective manner. Nowadays, the importance of power quality aspects has increased due to the booming developments in power-electronic devices and renewable energy resources under the umbrella of smart grids. Besides, the deregulation of the electricity market resulted in a competitive market in which multiple utility companies try to deliver the best products (generated electrical energy) for the customers who have the chance to choose the utility company that provides them with electrical energy with the highest quality level. In consequence, power quality will play an essential role in modern electrical power systems. However, there are also difficulties before wider applicability is possible for the power quality performance limits. One difficulty is that, to date, there is no single commonly approved definition of power quality because of the various power quality perspectives and phenomena [1]. As well, power quality has dissimilar interpretations for people in various electric entities. Some express power quality as the voltage quality, others express it as the current quality, and some practice power quality as the system reliability. Furthermore, IEEE Std. 1100 [2] defines power quality as "the concept of powering and grounding sensitive electronic equipment in a manner suitable for the equipment." One can say that everyone describes it from his own perspective.

IntechOpen

On one side, voltage quality focuses on variations of voltage from its ideal waveform (i.e. characterized by a sine wave of constant magnitude and frequency), while current quality is concerned with the deviation of the current from the ideal sinusoidal waveform. On the other side, discrimination of power quality as a voltage quality or current quality is an ambiguous way of thinking as a deviation in voltage can result in a deviation in current and vice versa. Thus, in order to keep generality, and as the power is mathematically the voltage times current, power quality should be the combination of both voltage and current qualities [3] and is signified by a set of electrical limitations (reference boundaries/margins) that enable an equipment to operate in its planned manner without major operating losses [4, 5] to long live as possible.

2. Disturbances

All electrical equipment may fail or malfunction when come across power quality disturbances, depending on the severity of the disturbance. It is essential for engineers, technicians, manufacturers, and power system operators to well understand and face the several power quality disturbances.

Power quality issues include voltage variations (dips, interruptions, flicker, etc.), transients (surges, lightning, and switching events), and grounding issues. **Figure 1** summarizes the common power quality problems.

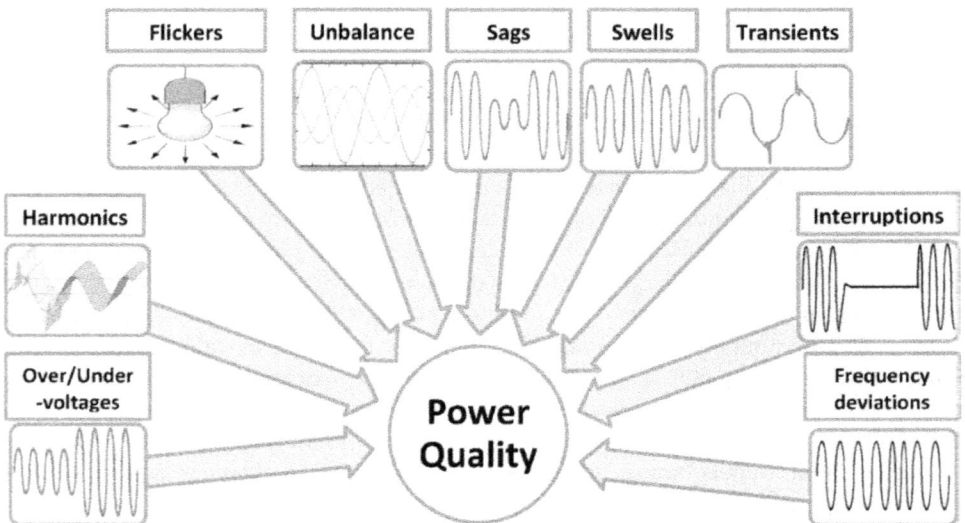

Figure 1. The common power quality problems.

To generalize, power quality issues cover many power system problems like impulsive and oscillatory transients, different types of interruptions, voltage sags and swells, imbalance, under and over voltages, notching, noise, harmonics and interharmonics, voltage fluctuations and flickers, and power frequency variations [6]. In the following sections, these power quality problems are presented.

2.1. Over voltages and under voltages

Over voltages are defined as any voltage greater than the equipment nominal operating voltage when the equipment is specified to operate at for a time period that exceeds 1 min. While, the under voltage can be defined as any voltage below the nominal operating voltage of the equipment for a time period that exceeds 1 min.

Over-voltage phenomenon has many causes in power system networks such as sudden changes in the system operating settings, abrupt load rejection, series/parallel harmonic resonance cases, sudden line-to-ground faults, improper earthing schemes, poor voltage regulation throughout the system, and overcompensation of the reactive power support provided by capacitor banks. Under voltages can result from improper power cables sizing, long feeder routes with high loading capacities, and large motor starting conditions.

Over voltage has a serious impact on electrical equipment and power systems as it stresses the equipment's insulation and may damage it, in addition to protective devices tripping because of dielectric failure. Also, over voltage may lead to flashover between line and ground at the weakest point in the system and can cause breakdown of the equipment insulation. On the other hand, under voltage causes an increase in the system losses and results in voltage stability problems. Also, different operational problems may arise due to under voltages such as motor starting problems and protection relay tripping [7].

2.2. Voltage flickers

Voltage flickers are defined as a continuous rapid variation of input supply voltage sustained for an appropriate period to enable visual recognition of a variation in electric light intensity. Flicker is a power quality problem in which the magnitude of the voltage or frequency changes at a rate that is to be noticeable to the human eye [6]. The main causes of the voltage flicker are the loads that draw large starting currents during initial energization such as elevators, arc furnaces, and arc welders. If load starting cases are rapidly repeated, then light flicker effects can be quite noticeable. The severity of voltage flickers is measured using short-term and long-term flicker severity terms, where an expected flicker severity over a short duration (typically 10 minutes) is known as P_{st}, and that evaluated over a long duration (typically 2 hours) is known as P_{lt}. Thus, P_{lt} is a combination of 12 P_{st} values.

$$P_{st} = \sqrt{(0.0314 \times P_a) + (0.0525 \times P_b) + (0.0657 \times P_c) + (0.28 \times P_d) + (0.08 \times P_e)} \qquad (1)$$

where P_a, P_b, P_c, P_d, and P_e are the surpassed flicker levels during 0.1, 1, 3, 10, and 50% of the surveillance period. By definition, a value of one for P_{st} expresses a visible disturbance, a level of optical severity at which 50% of persons might sense a flicker in a 60-W incandescent lamp. Excessive light flicker can cause a severe headache and can lead to the so-called 'sick building syndrome.'

2.3. Voltage unbalance

Voltage unbalance problem is an important power quality issue that can be defined as "a condition in a three-phase system in which the root-mean-square (rms) magnitudes and/or phase angles of the fundamental components of the phase voltages are not all equal" [7]. The principal reason of voltage unbalance in a system is the unbalanced loads among the three phases of the network. This asymmetric loading causes unequal phase currents to flow through the electrical network, and causes unsymmetrical voltage drop on system feeders [8]. Voltage unbalances result in additional power losses in the system and cause more losses in electric motors, so that it cannot be completely loaded up to its nominal power. In addition, excessive voltage unbalances can lead to protection system tripping and cause electrical supply interruption.

The IEEE 112 [9] defines the voltage unbalance using a factor called the phase voltage unbalance rate (PVUR), is given in (2), where V_{dev} expresses the phase voltage variation from the average line voltage ($V_{average}$) [10].

$$PVUR = \frac{(V_{dev})Max}{V_{average}} \times 100 \qquad (2)$$

2.4. Voltage sags

Voltage sags or (American English says sag while British English says dip) According to the IEEE-1159 [11], voltage sag is defined as "a reduction in the rms voltage from 0.1 to 0.9 per unit (pu) for a period of 0.5 cycles to 1 minute." Voltage sag can be categorized into three types, according to its time duration, to instantaneous, momentary, and temporary sag [12].

Voltage sag results from sudden system faults and switching events of large loads having excessive starting currents such as large motors. Voltage sags impact on sensitive electrical devices such as personal computers and communication equipment, as well as excessive sag events may cause loss of data and nuisance operation of protection devices. In addition, programmed industrial processes such as paper-making industries, chip-making machinery, etc. can suffer from power supply shutdown in case of severe voltage sags.

Voltage sags can be calculated using various formulas. For example, Detroit Edison's sag score (SS) method defines the "sag score" from the amplitudes of the three-phase voltages. A larger SS indicates the more the severity of the event [13].

$$SS = 1 - \left(\frac{V_A + V_B + V_C}{3}\right) \qquad (3)$$

2.5. Voltage swells

Voltage swell can be defined as a rise in the root mean square (rms) voltage for periods that range from 0.5 cycles to 1 minute. Swells are usually produced by electric faults (single line-to-ground), upstream supply failures, heavy load rejection events, and switching off shunt capacitor banks. Voltage swell is categorized, according to time duration, into three types: instantaneous, momentary, and temporary swells. In addition, voltage sags and swells are produced when loads are shifted from one supply source to another such as the transfer from the utility source to the standby emergency generator during a loss of the normal utility power source [14].

Voltage swell has harmful effects on electrical power system operation as it leads to aging of electrical connections, flickering of lights, semiconductor damage in power-electronic devices, and insulation deterioration of the equipment.

2.6. Transients

In general, most power quality problems are thought as transient events if they exist for a short duration. Impulsive and oscillatory are the main categories of transients. They are briefed as follows:

A. Impulsive transients

Impulsive transients are abrupt high magnitude actions that cause voltage and/or current levels to rise in either a positive or a negative direction for a very short period fewer than 50 ns.

3. Oscillatory transients

An oscillatory transient is an abrupt variation in the steady-state voltage, current, or both, fluctuating at the natural frequency of the system at both the positive and negative directions.

Events causing transients occur from different reasons such as lightning strikes, poor grounding system, electrostatic discharge, inductive load switching, and fault clearance. Transients may lead to probable data loss in computers, malfunction of electronic equipment, and microprocessor-based protection relays.

2.7. Interruptions

Interruption is a randomly event that occurs with zero-magnitude voltage or current for a particular time period, where the magnitude of voltage or current is less than 0.1 pu. It is classified in terms of duration and standards as follows:

A. Classification according to prior planning

According to EN 50160 [15], the electrical interruptions can be categorized into two types, namely, pre-organized interruptions at which the customers are informed (planned interruptions) and accidental interruptions at which sudden failure of equipment or transient fault take place and it may take a long time to restore the electrical supply. This may be long interruption or short interruption based on the fault.

B. Classification according to interruption duration

According to IEEE 1250 [16], the electrical interruptions can be categorized into four types according to the duration of the interruption as present in **Table 1**.

Momentary interruptions may cause a complete loss of voltage, while sustained interruptions are generally noticed in case of permanent short-circuit faults.

2.8. Frequency deviation

The fundamental frequency varies from its rated (50 or 60 Hz) value. This frequency deviation is infrequent in stable and stiff interconnected power system networks. However, it can be noticed in weak power systems fed from local generators especially during sudden load application or rejection conditions.The operating frequency range should be kept within ±1% the rated frequency for 95% of week and -6%/+4% for 100% of week [15, 16]. The ratio of frequency deviation (RFD) can be defined as follows:

$$RFD = \frac{|f_m - f_r|}{f_r} \times 100 \qquad (4)$$

where f_m is the measured frequency which is time-varying quantity and f_r is the rated system frequency.

2.9. Power system harmonics

Most of today's power system waves are distorted. By definition, "any periodically distorted waveform can be represented as a sum of pure sine waves in which the frequency of each sinusoid is an integer multiple of the fundamental frequency of the distorted wave. This multiple is called the harmonic of the fundamental." The sum of sinusoids is referred to as a "Fourier series."

In the last years, all have focused on power system harmonic distortion, because it has adverse impacts on both the utility and consumers, alike. Sometimes, when the terminology of power

Type of interruption	Duration starts at	Duration ends at
Instantaneous	0.5 cycles	30 cycles
Momentary	30 cycles	2 s
Temporary	2 s	2 min
Sustained	Longer than 2 min	

Table 1. Electrical interruptions categorized based on their durations.

quality arises, some people routinely predict that the issue is related to power system harmonic distortion. In the past, the terms of power quality and power system harmonics have been incorrectly interchanged.

2.9.1. Harmonics sources

At present, as a consequence of the extensive use of power electronic-based components in all power system applications, most of today's loads are nonlinear. To generalize, three categories can be recognized as primary sources of harmonics in power systems [6]. They are given as follows:

- Magnetic core-based equipment as electric motors, power transformers, and generators.

- Arc and induction, and arc welders.

- Power electronic-based equipment.

On one hand, if the power system is characterized by series and shunt elements; thus, the nonlinearities exist in the system are mainly introduced by the shunt elements, such as loads. On the other hand, a series impedance of the power delivery system (impedance between the source and the load) is particularly linear, that is, short circuit or Thevenin impedance of a system. Even within a power transformer, the shunt branch (magnetizing impedance) of the standard T model is the source of harmonics, while the series leakage impedance is considered as a linear element.

Today, the most prevailing harmonic sources are:

- Converters (rectifiers and inverters).

- Switch-mode power supplies.

- The different forms of pulse modulation which are employed in active power and voltage control in transmission circuits.

- High-frequency converters needed for induction heating.

- Thyristor controlled reactors.

- Rectifiers and inverters of grid-connected solar photovoltaic cells and windmills.

- Magnetizing currents of the transformers.

- Excitation currents of the rotating machines.

- Flexible AC transmission systems (FACTS).

- Uninterruptible power supplies.

- Rectifiers and inverters of HVDC systems.

- Static power converters using thyristor to control speed and torque of variable speed drives.

- Controlled arc welders, controlled furnaces, and ovens.

- Induction motors working in the saturation region.

- Electrolysis loads (aluminum smelters and battery-charging plants).

- Ballasts in high-power fluorescent discharge lamps.

2.9.2. Harmonics effects

Impact on harmonics can range from degradation of performance of equipment to its serious failure. The effects of power system harmonics can be clustered into two broad groups: as effects on power system networks and equipment and effects on telecommunication systems.

The most common consequences on the different sectors of an electrical system are summarized below [17].

- Excessive energy losses due to the high nonsinusoidal currents, thus leading to high electricity bills.

- The presence of current in the neutral wire with additional losses. An overheating problem may occur.

- Equipment failure, standstill of motors, overloading of conductors, blowing of fuses, and blackouts of lamps.

- Errors in metering of energy consumption.

- Interference with telecommunication systems and networks.

- Data loss in data-transmission systems.

- Malfunction of control and protection system performance.

- Series and parallel harmonic resonance, which may cause system component damage, equipment failure, and service interruption.

- Harmonic instability which leads to the damage of generator shafts.

- Audible noise in transformers, rotating machines, and motor vibrations.

- Computer and programmable logic controllers' lockups and in correct operation.

- Malfunctioning of voltage and generator regulators with frequent maintenance issues.

- Premature aging of equipment.

- UPS sizing issues.

- Worsening of loads' power factor with its adverse consequences and utility's imposed penalties.

3. Solutions

A thorough understanding of electrical system related problems is helpful to implement good power conditioners and custom power devices to enhance the power quality. Today, it is assumed that the most of our electrical loads become nonlinear in nature. Generally, power factor improvement and other power quality-based equipment are the two main groups of solutions that can enhance the power quality performance in a system, thus:

A. Power factor improvement equipment [17–24]

• Power factor correction capacitors.

• Harmonic filters, especially passive filters.

These solutions certainly guarantee energy bill savings from reduction of low power factor penalties, not power or energy savings [24].

B. Other power quality-based equipment [17, 24–27]

• Inline reactors or chokes.

• Harmonic mitigating and K-factor transformers.

• Neutral blocking filter.

• Negative sequence current reduction.

• Passive, active, and hybrid filters.

• Surge protection.

• Soft starters.

• Zigzag reactors.

• Conservation voltage reduction.

• Green plug filters, FACTS, and D-FACTS.

• Multiple pulse converters.

These solutions can enhance the power quality but with no real savings [24].

Each power quality solution has its own merits and drawbacks at different circumstances. Consequently, selection of a precise solution to solve a power quality problem necessitates familiarity with the different technologies to ensure that it is the proper techno-economic solution for an application.

Besides, as the grids transition toward low-carbon technologies, the use of power electronics becomes widespread. Also, renewable sources may introduce harmonic distortions which may adversely affect consumer equipment, but also monitoring and controlling devices that maintain the operational status of the grids themselves, which can lead to large-scale black-outs and significant losses in power networks. Therefore, it is imperative that novel solutions be sought to enable networks to cope with future developments.

Finally, power quality issues cover many power system problems such as under and over voltages, voltage sags and swells, transients (impulsive and oscillatory), interruptions, volt-age unbalance, harmonics and interharmonics, voltage fluctuations and flickers, and power frequency variations. In this introductory chapter, a quick brief on power quality concepts and issues are presented.

Author details

Ahmed F. Zobaa[1]*, Shady H.E. Abdel Aleem[2] and Murat E. Balci[3]

*Address all correspondence to: azobaa@ieee.org

1 College of Engineering, Design and Physical Sciences, Brunel University London, Uxbridge, United Kingdom

2 15th of May Higher Institute of Engineering, Mathematical and Physical Sciences, Cairo, Egypt

3 Electrical and Electronics Engineering, Balikesir University, Balikesir, Turkey

References

[1] Elbasuony GS, Abdel Aleem SHE, Ibrahim AM, Sharaf AM. A unified index for power quality evaluation in distributed generation systems. Energy. Apr. 2018;**149**:607-622

[2] IEEE Std 1159-2009. IEEE Recommended Practice for Monitoring Electric Power Quality USA: IEEE; 2009, pp.c1-81

[3] Bollen M, Gu I. Signal Processing of Power Quality Disturbances. USA: Wiley-IEEE Press; 2006

[4] Dixit JB, Yadav A. Electrical Power Quality. New Delhi, India: Laxmi Publications Ltd.; 2010

[5] Aleem SHEA, Balci ME, Zobaa AF, Sakar S. Optimal passive filter design for effective utilization of cables and transformers under non-sinusoidal conditions. In: 2014 16th International Conference on Harmonics and Quality of Power (ICHQP); Bucharest: IEEE; 2014. pp. 626-630

[6] Dugan RC, Granaghan MFM, Santoso S, Beaty HW. Electric Power Systems Quality. 2nd ed. New York: McGraw-Hill; 2002

[7] Kurt MS, Balci ME, Abdel Aleem SHE. Algorithm for estimating derating of induction motors supplied with under/over unbalanced voltages using response surface methodology. Journal of Engineering. Dec. 2017;**2017**(12):627-633(6)

[8] Baggini A. Handbook of Power Quality. United States: John Wiley & Sons; 2008

[9] IEEE Standard 112-1991. IEEE Standard Test Procedure for Polyphase Induction Motors and Generators. USA: IEEE; 1991

[10] Saeed AM et al. Power conditioning using dynamic voltage restorers under different voltage sag types. Journal of Advanced Research. Jan. 2016;**7**(1):95-103

[11] Balasubramaniam PM, Prabha SU. Power quality issues, solutions and standards: A technology review. Journal of Applied Science and Engineering. 2015;**18**:371-380

[12] Zobaa AF, Abdel Aleem SHE, editors. Power Quality in Future Electrical Power Systems Energy Engineering. United Kingdom: IET Digital Library; 2017

[13] Polycarpou A. Power quality and voltage sag indices in electrical power systems. In: Romero G, editor. Electrical Generation and Distribution Systems and Power Quality Disturbances. Croatia: InTech; 2011. p. 140-160. DOI: 10.5772/18181

[14] Sankaran C. Power Quality. United States: CRC Press; 2001

[15] BS EN 50160:2010+A1:2015. Voltage characteristics of electricity supplied by public electricity networks. UK: BSI; 2010

[16] IEEE Standard 1250-2011. IEEE Guide for Identifying and Improving Voltage Quality in Power Systems. USA: IEEE; 2011, pp.1-70

[17] Carnovale DJ. Applying Harmonic Solutions to Commercial and Industrial Power Systems. Moon Township, PA, Boston: Eaton Cutler-Hammer; 2003

[18] Aziz MMA, El-Zahab E-D, Ibrahim AM, Zobaa AF. Practical considerations regarding power factor for nonlinear loads. IEEE Transactions on Power Delivery. 2004;**19**(1):337-341

[19] Balci ME, Abdel Aleem SHE, Zobaa AF, Sakar S. An algorithm for optimal sizing of the capacitor banks under nonsinusoidal and unbalanced conditions. Recent Advances in Electrical & Electronic Engineering. 2014;**7**(2)

[20] Aleem SHEA, Zobaa AF, Balci ME. Optimal resonance-free third-order high-pass filters based on minimization of the total cost of the filters using crow search algorithm. Electric Power Systems Research. 2017;**151**(C):381-394

[21] Independent pricing and regulatory tribunal of new south wales, method guide power factor correction energy savings formula: deemed energy savings method. Energy Savings Scheme. Jan. 2015. Available from: http://www.ess.nsw.gov.au/files/353708d1-ab17-4aa4-96a5-a41b0103d03f/Method_Guide_-_Power_Factor_Correction_-_V20.pdf

[22] Aleem SHEA, Elmathana MT, Zobaa AF. Different design approaches of shunt passive harmonic filters based on IEEE Std. 519-1992 and IEEE Std. 18-2002. Recent Advances in Electrical & Electronic Engineering. 2013;**6**(1):68-75

[23] Mohamed IF, Aleem SHEA, Ibrahim AM, Zobaa AF. Optimal sizing of C-type passive filters under non-sinusoidal conditions. Energy Technology & Policy. 2014;1(1):35-41 DOI: 10.1080/23317000.2014.969453

[24] Carnovale DJ, Hronek TJ. Power quality solutions and energy savings—What is real? Energy Engineering: Journal of the Association of Energy Engineers. 2009;106(3):26-50

[25] Heetun KZ, Abdel Aleem SHE, Zobaa AF, Aleem SHEA, Zobaa AF. Voltage stability analysis of grid-connected wind farms with FACTS: Static and dynamic analysis. Energy and Policy Research. Jan. 2016;3(1):1-12

[26] Gandoman FH, Sharaf AM, et al. Distributed FACTS stabilization scheme for efficient utilization of distributed wind energy systems. International Transactions on Electrical Energy Systems. 2017;27(11):e2391

[27] Smart ways to cut down the influence of harmonics. Available from: http://electrical-engineering-portal.com/smart-ways-to-cut-down-the-influence-of-harmonics

Sequential Selective Harmonic Elimination and Outphasing Amplitude Control for the Modular Multilevel Converters Operating with the Fundamental Frequency

Alexey Tyshko

Additional information is available at the end of the chapter

Http://dx.doi.org/10.5772/intechopen.72198

Abstract

With the growing use of DC voltage for power transmission (HVDC) and DC links for efficient AC motor drives, the R&D efforts are directed to the increase of DC/AC converter's efficiency and reliability. Commonly used DC/AC converters, based on the carrier-frequency pulse-width modulation (PWM) to form a sinusoidal output voltage with a low level of higher harmonics, have switching time and switching loss issues. The use of multimodule multilevel converters (MMC), operating with the fundamental switching frequency and phase-shift control to form the ladder-style output voltage, reduces switching losses to minimum while keeping the low level of higher harmonics in the output voltage. The discussed sequential harmonic elimination method for MMC, using identical power modules operating with 50% duty cycle and fundamental frequency, is based on the combination of the multiple fixed phase shifts to form a ladder-style sinusoidal voltage with low total harmonic distortion (THD) and symmetrical variable phase shifts to control the output voltage amplitude. The principles of the sequential selective harmonic elimination for MMC topology and amplitude control are described with two examples. The first example is the industrial-frequency DC/AC converter complying with THD requirements of IEEE 519 2014 standard without the output filter. The second example is a high-frequency converter, used as a transmitter, loaded with the resonant antenna, where the evaluation criteria are decreasing of the transmitter losses and increasing of the reliability or life expectancy at elevated temperature.

Keywords: amplitude control, selective harmonic elimination, staircase modulation, Chireix-Doherty amplitude modulation, DC/AC converter, high temperature, multimodule multilevel converter (MMC), multivector control, outphasing, phase-shift modulation, reliability, life expectancy

1. Introduction

The conversion of DC voltage into sinusoidal AC voltage at power levels from kilowatts to megawatts with low power losses and low higher harmonics in the output voltage is a common task for modern power engineering. Multimodule multilevel converter is the best approach to generate the high-power sinusoidal voltage from HVDC bus for electrical grid consumers propulsion electrical motor drives, etc. High-efficiency switch-mode modules, used to synthesize sinusoidal output voltage, may operate at the fundamental frequency of the sine voltage required for the load, or at higher frequencies (the carrier frequency) using the pulse-width modulation to reduce higher harmonics of the fundamental frequency. In the last case, the output filters, required for reducing total harmonic distortion (THD) of the output voltage to the acceptable level, are significantly smaller [1–20].

The biggest problem with the phase-shift pulse-width modulation, providing the highest quality of the output sinusoidal voltage with minimum switching losses, is its control methodology, which requires complicated calculation of the necessary phase shifts in real time [8, 21, 22]

In this paper a simple method of the sequential selective harmonic elimination and amplitude control is discussed. It is based on the combination of the fixed precalculated phase shifts, delays for harmonic elimination and variable phase shift for amplitude control. Application of this method is illustrated using two examples—the industrial-frequency DC/AC converter and the high-frequency converter used as a transmitter for the nuclear magnetic resonance (NMR) oil/gas well logging tool, operating in harsh conditions. LTspice was used for simulation in time and frequency domains. A simple expression is provided for the resulting THD vs. the number of eliminated harmonics to comply with industrial grid voltage of THD standards without the output filter. For the NMR transmitter, decreasing of conductive losses due to the harmonic elimination reduces operating temperature and increases the reliability Improvement of the life expectancy is calculated according to the Arrhenius equation for three transmitter cases with the same number of switches but with different harmonic contents.

2. Full-bridge module operation and spectrum of the output voltage

The building block or module for multimodule multilevel converter (MMC) is a full-bridge DC/AC converter utilizing maximum voltage and current ratings of the power switches S1–S4 (**Figure 1**), powered from bus $V_{o'}$ producing rectangular voltage pulses with 50% duty cycle

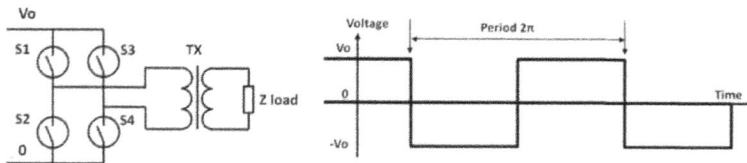

Figure 1. Full-bridge stage and output voltage waveform.

for maximum output power. The bridge load Z is connected directly to the bridge outputs or via the output transformer TX. For the industrial frequency 50 Hz–60 Hz and other low-frequency high-power applications, fully controlled thyristors are the best choice, while for the frequency range over few kilohertz, IGBTs are the preferred ones. Operation in the frequency range over 100 kHz requires fast-switching power MOSFETs. To simplify analysis of the following circuits, the switches are assumed to be ideal and have zero-switching time and zero internal losses.

The Fourier analysis provides the expression for the full-bridge symmetrical 50% duty cycle output voltage $V_{out(t)}$ (**Figure 1**) as the sum of only odd harmonics V_n (n = 1, 3, 5, 7, etc.):

$$V_{out(t)} = \frac{4 V_0}{\pi} \sum_{n=1} \frac{\cos n\omega t}{n} \tag{1}$$

where n is the harmonic number (only odd harmonics 1, 3, 5, etc.), ω is the angular frequency, V_0 is the full-bridge inverter DC bus voltage and t is time.

Each harmonic n has its amplitude V_n decreasing with the harmonic number n:

$$V_n = \frac{4 V_0}{\pi n} \tag{2}$$

Spectrum of the bridge output voltage with amplitude of 1 V and frequency of 1 kHz is shown on **Figure 2**. The vertical axis represents the RMS values of each harmonic starting with the first one equal to 0.9Vrms (or 1.273 V peak value). Horizontal axis is frequency.

Converter output current $I_{out(t)}$ is a combination of the fundamental harmonic and higher harmonics, each of them being a product of the harmonic voltage $V_{n(t)}$ and load admittance Y_n for this harmonic:

$$I_{out(t)} = \sum_{n=1}^{\infty} V_{n(t)} Y_n \tag{3}$$

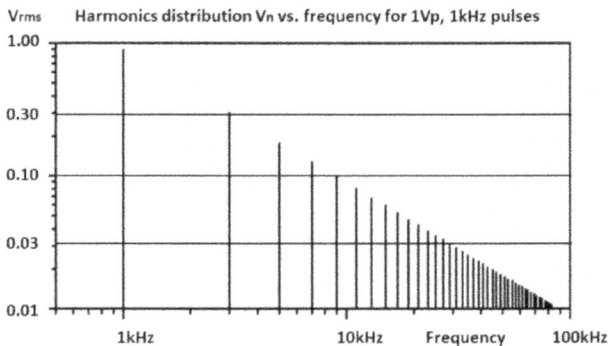

Figure 2. Spectrum of the 1 kHz 50% duty cycle signal.

Several load types such as resistive, inductive, capacitive and resonant ones have different current vs. frequency characteristics as shown in **Figure 3**, which is obtained in LTspice environment under 1 V sinusoidal test signal.

Only resistive load current replicates the spectrum of the input voltage. Inductive load decreases high-frequency current components, but capacitive and resonant loads significantly increase relative values of the high-frequency current harmonics compared to the spectrum of the applied voltage. Voltage harmonics and resulting currents affect both load and voltage sources (converter) in different ways. Excessive current harmonics increase power losses and create electrical noise (EMI) affecting electronic equipment.

Maximum voltage harmonic content for the industrial AC lines is regulated by IEEE 519 2014 standard [23, 24]. Limits for total harmonic distortion (THD) and maximum amplitude of the highest harmonic are provided in **Table 1**.

THD and individual harmonic maximum values are different for different line voltages. The power distributor should keep total harmonic distortion (THD) for voltages <1 kV under 8% and individual harmonic value less than 5% of the fundamental one at the point of consumer connection (PCC). In the process of conversion of HVDC bus voltage into lower-level AC, the switch-mode converters create higher harmonics as unwanted byproduct. For full-bridge DC/AC converter output voltage spectrum (**Figure 2**) of THD is 0.483 or 48.3% [25]. To comply with THD limits, the simple DC/AC converters include the output filters reducing higher harmonics to the acceptable level. Those filters introduce additional losses and have significant size, weight and cost especially if the filter has to remove harmonics starting with the third one,

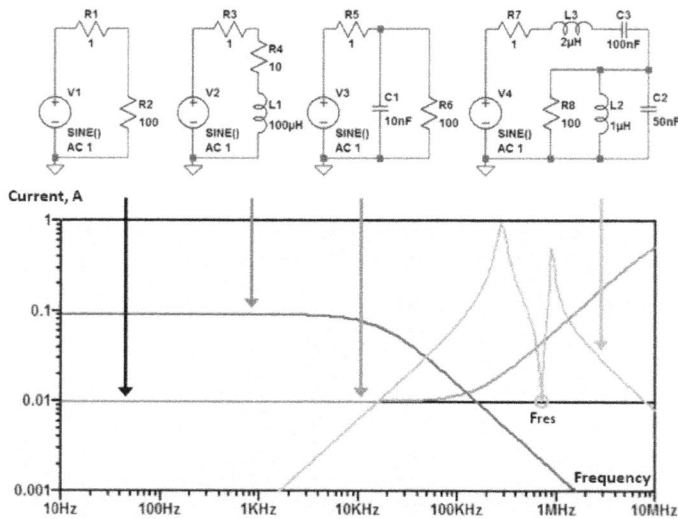

Figure 3. Load current vs. frequency for different loads.

Bus voltage (V) at PCC	Individual harmonic (%)	Total harmonic distortion (THD) (%)
V ≤ 1.0 kV	5.0	8.0
1 kV < V ≤ 69 kV	3.0	5.0
69 kV < V ≤ 161 kV	1.5	2.5
151 kV < V	1.0	1.5

High-voltage systems can have up to 2% THD where the cause is an HVDC terminal where effects will have attenuated at the point in the network where future users may be connected.

Table 1. Voltage distortion limits.

which is 150 Hz and 180 Hz for the EU and USA, respectively. Eliminating the most powerful higher harmonics from the output voltage in the process of DC to AC conversion and reducing the highest-frequency harmonic leftovers with a simple output filters are the most efficient ways to comply with THD standard.

For the high-frequency converters operating as transmitter with the resonant loads at elevated temperature, the output current's higher harmonics cause additional heating, which results in the reliability problems. In this case the effectiveness of the harmonic elimination is reducing the power component temperature and increasing the converter life expectancy.

3. Multimodule converters and synthesis of the quasi-sinusoidal output voltage

Multimodule multilevel converters (**Figure 4**) have their outputs connected in series to produce the so-called modified sinusoidal voltage or ladder-style voltage (**Figure 5**). DC inputs may be connected in parallel with the transformer combining the output voltages or in series for HVDC

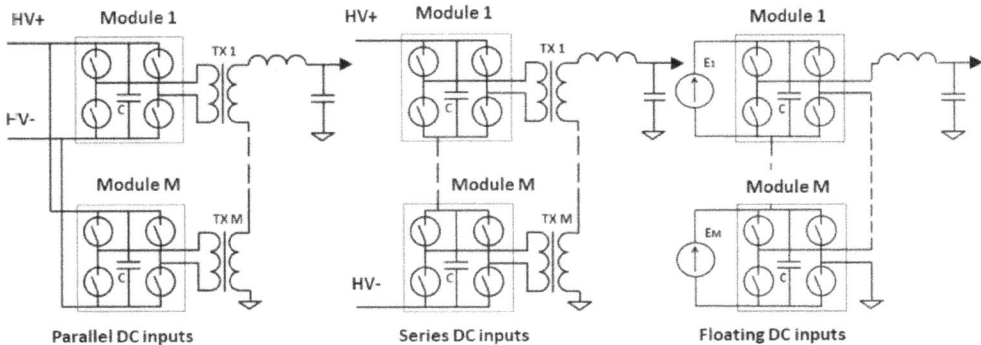

Figure 4. Multimodule converters with different DC line feeds.

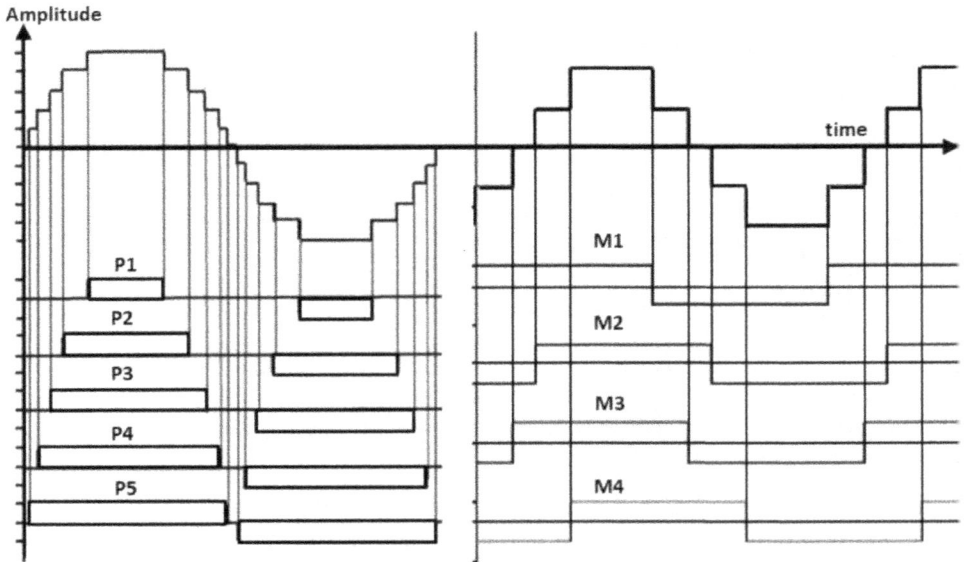

Figure 5. Forming ladder-style voltage using PWM (left) and phase shift only (right).

converters or be floating, for example, powered from the photovoltaic batteries, depending upon the application. Each module operates with high efficiency producing rectangular pulses with controlled timing. The control algorithm for timing calculations is a subject of this analysis.

Commonly used control algorithms are based on the pulse-width modulated signals to control multiple modules. The module output voltages are added in series to form the ladder-style voltage with optimized width of each pulse to eliminate harmonics and to regulate output voltage as shown in **Figure 5** (left). The discussed method was developed to operate the identical modules, producing 50% duty cycle pulses with the fundamental frequency and equal amplitude, combining their outputs in series and controlled by the phase shift only (**Figure 5**, right).

Equal ON and OFF time operation of all power switches has the following advantages:

a) Equal conductive losses

b) Guaranteed time to reset snubbers (if used to reduce switching losses)

c) Guaranteed time to build up the lagging current for soft switching (if needed)

d) Guaranteed time to recharge gate drivers (if needed)

4. Harmonic elimination based on the phase shift

Minimal configuration of the DC/AC multimodule converter includes two basic full-bridge modules with the output voltages connected in series. Their DC inputs may be connected in parallel to the same bus V_0 or in series (this is used for HVDC lines to share input DC voltage according to the module maximum voltage rating).

To simplify analysis two identical modules are connected to the same DC bus, and their outputs are connected in series using two ideal output transformers TX1 and TX2 with transformer ratio 1:1 (**Figure 6**). The output voltages have identical amplitude and 50% duty cycle, and their relative position in time domain (delay or phase shift) is defined by the controller (not shown).

The relative phase shift (relative to the fundamental harmonic) equal to $\pi/3$ or 1/6 of the module operation period is shown in **Figure 7**. For the third harmonic, the relative phase shift is π, and combined output signal has the third harmonics subtracted or eliminated. All other odd harmonics like 9th, 15th, 21st, 27th, etc., which are multiples of three, are eliminated too.

Two identical periodic signals being combined with the phase shift Φ have eliminated harmonics n satisfying the following requirement:

$$\Phi = \frac{\pi}{n} \tag{4}$$

Two module output voltages shifted $\pi/3$ and combined output without 3rd, 9th, etc. harmonics are shown in **Figure 8**. When two signals with eliminated 3rd harmonic are added with phase shift $\pi/5$, their combined voltage has eliminated the 5th harmonic and also 15th, 25th, etc. To eliminate the fifth and seventh harmonics, this process should be repeated as shown in **Figure 9** where control signals to a set of eight modules are getting additional delay starting from sync pulse. This process may be continued to eliminate enough harmonics to comply with THD requirements or other special conditions [26–31].

Total delays (or phase shift) α_m per module m of a set of modules M (**Figure 9**) may be calculated as scalar product of matrix C formed by assigned to each module m binary numbers ($c = m-1$) of the modules and set Φ of harmonic canceling delays $\phi = \pi/n$:

$$\alpha_m = [c_m] \cdot [\varphi] \tag{5}$$

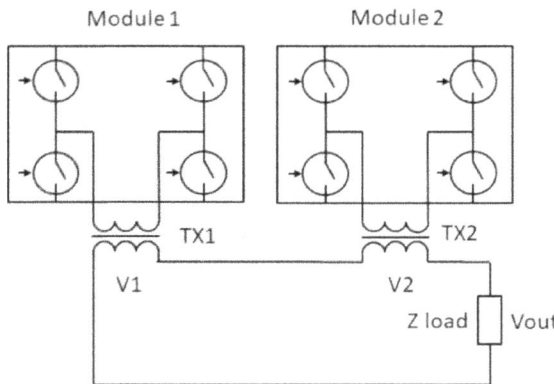

Figure 6. Two module converter.

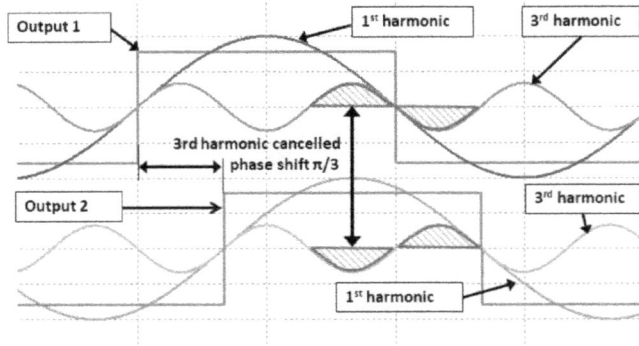

Figure 7. Third harmonic canceling.

Examples of delay/phase-shift calculation for the eight modules of MMC are provided in Table 2.

Number M of modules, needed for eliminating K harmonics

$$M = 2^K \qquad (6)$$

To find THD dependence on the sequential selective harmonic elimination using phase shift, seven harmonic elimination circuitries with different numbers of modules, marked A to G, were simulated. Example of circuitry topology marked as A, B, C and D, reference A full spectrum single module and with eliminated third (B-2 modules), third and fifth (C-4 modules) and third, fifth and seventh (D-8 modules) harmonics are provided in **Figure 10**. More complicated circuits (E-16 modules, F-32 modules and G-64 modules) were also simulated.

THD for each simulated case (A to G) was calculated based on the RMS value of the simulated output ladder-style voltage V_{out} and RMS value of the fundamental harmonic V_1 [25]:

$$THD = \frac{\sqrt{V_{out}^2 - V_1^2}}{V_1} \qquad (7)$$

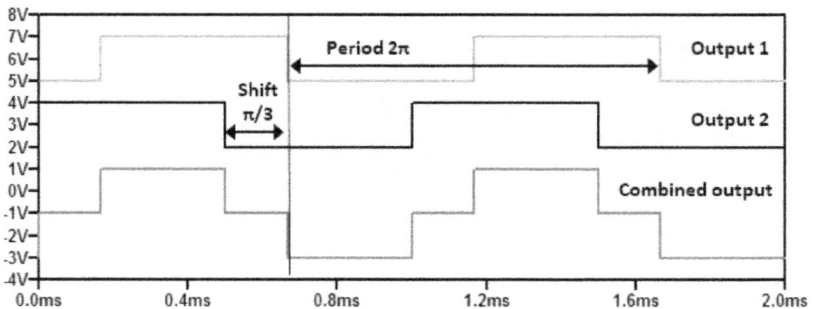

Figure 8. Forming ladder voltage with two modules with canceled the third harmonic.

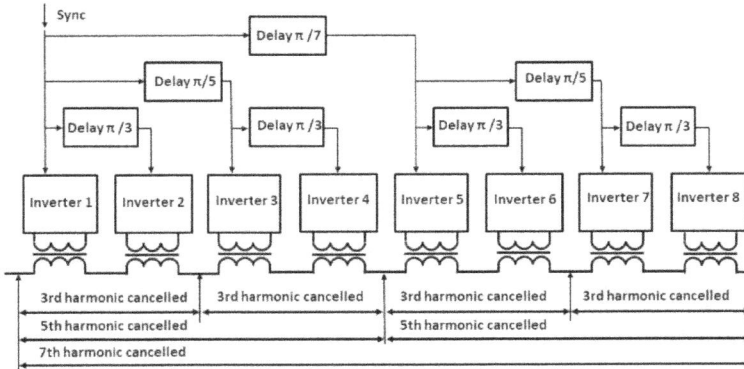

Figure 9. Topology for canceling the third, fifth and seventh harmonics.

The drawback of any kind of modulation is decreasing of the resulting fundamental harmonic V_1. **Table 3** provides not only THD and maximum value of the biggest voltage harmonic left after multiple harmonic eliminations but also a change of the fundamental harmonic compared to expected value in case of all phase shifts, which were zero, and combined output voltage replicates 50% duty cycle output voltage of the single module multiplied by the number of modules.

THD value vs. the number of eliminated harmonics starting from the third from Cases A to G is presented in **Figure 11**. The relative amplitude of the fundamental harmonic $V_{1,k}$ after the first k higher harmonics are eliminated is calculated as

$$V_{1,k} = \frac{4\,V_0}{\pi} \prod_{n=3}^{k} Sin\left\{ \frac{\pi}{2}\left(1-\frac{1}{n}\right) \right\} \tag{8}$$

Module Number (m)	Module's assigned binary code ($c_m = m-1$)	nth harmonic phase shift in module ($\varphi_n = \pi/n$)			Total module phase shift ($\alpha_m = c_m \cdot \Phi$)	Total module phase shift (α_m)
		n = 7	n = 5	n = 3		
1	0 0 0	0	0	0	0	0
2	0 0 1	0	0	φ_3	φ_3	$\pi/3$
3	0 1 0	0	φ_5	0	φ_5	$\pi/5$
4	0 1 1	0	φ_5	φ_3	$\varphi_5 + \varphi_3$	$\pi/5 + \pi/3$
5	1 0 0	φ_7	0	0	φ_7	$\pi/7$
6	1 0 1	φ_7	0	φ_3	$\varphi_7 + \varphi_3$	$\pi/7 + \pi/3$
7	1 1 0	φ_7	φ_5	0	$\varphi_7 + \varphi_5$	$\pi/7 + \pi/5$
8	1 1 1	φ_7	φ_5	φ_3	$\varphi_7 + \varphi_5 + \varphi_3$	$\pi/7 + \pi/5+ \pi/3$

Table 2. Phase shift per module calculation.

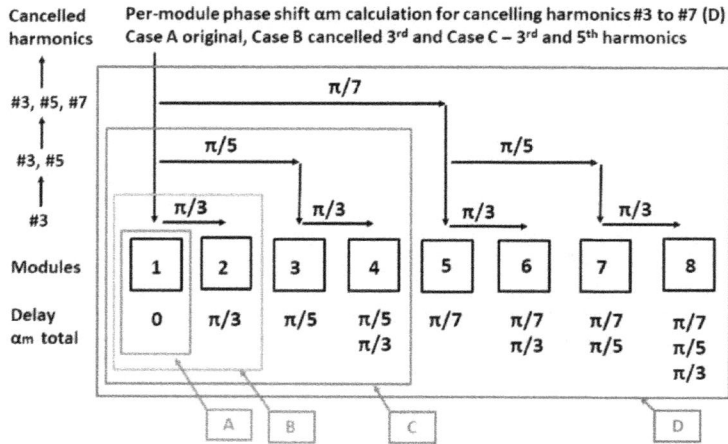

Figure 10. Harmonic elimination topology and phase-shift control.

Multimodule converter with eliminated the first four harmonics (Case E, 16 modules) complies with IEEE 519 2014 [23] requirements for voltage <1 kV, with five eliminated harmonics (Case F) for 1 kV to 69 kV and with six eliminated harmonics (Case G) from 69 kV to 161 kV without using any output power line filters. Eq. (9) may be useful for approximate calculation of THD of the MMC output voltage after canceling the first k harmonics using this method:

$$THD_k \approx 0.5 * e^{-0.5k} \tag{9}$$

Configuration	A	B	C	D	E	F	G
Number of modules	1	2	4	8	16	32	64
Harmonics canceled +	0	3	5	7	11	13	17
Phase shift per next module set	0	$\pi/3$	$\pi/5$	$\pi/7$	$\pi/11$	$\pi/13$	$\pi/17$
Relative change of the first harmonic	1.0000	0.8660	0.9511	0.9749	0.9898	0.9927	0.9957
Combined output value (Vrms)	1.0000	0.8165	0.7528	0.7278	0.7167	0.7108	0.7076
First harmonic k value V_1 (Vrms)	0.9003	0.7798	0.7415	0.7229	0.7156	0.7103	0.7074
THD	0.4834	0.3103	0.1752	0.1166	0.0575	0.0360	0.0238
Max higher harmonic number	3	5	7	11	17	29	29
Max harmonic value (Vrms)	0.3001	0.1559	0.0658	0.0526	0.0160	0.0127	0.0113
Max harmonic relative to the first harmonic	0.3333	0.1999	0.0887	0.0728	0.0224	0.0179	0.0160

Table 3. Simulation results for voltage THD and maximum high harmonic left.

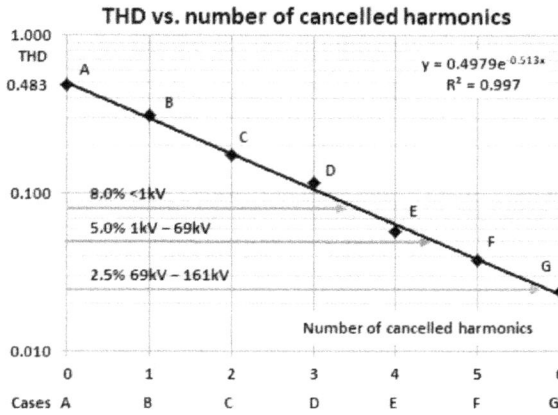

Figure 11. Voltage THD vs. the number of canceled harmonics starting with the third one.

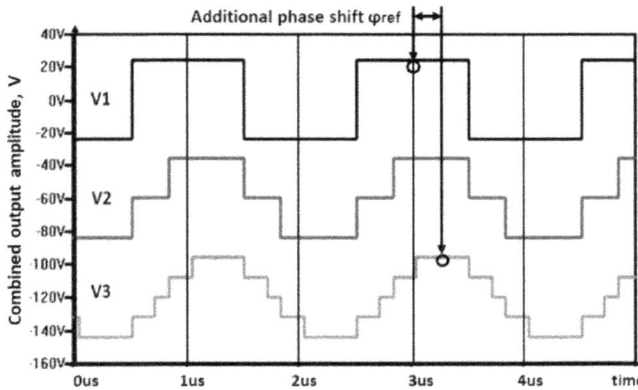

Figure 12. Buildup of the output voltage additional phase shift (case C vs. case a).

Due to the buildup of the module output voltage phase shifts, the resulting quasi-sinusoidal voltage fundamental harmonic is shifted to φ_{ref} as shown in **Figure 12** which should be taken into consideration and if necessary added to the carrier signal.

The additional phase-shift value φ_{ref} is calculated for MMC with M modules as

$$\varphi_{ref} = \frac{1}{2}\sum_{m}^{M} \alpha_{m} \tag{10}$$

5. Amplitude control

MMC with fixed phase shifts produces low THD sinusoidal output voltage with the amplitude proportion to the DC bus voltage. To regulate the output voltage amplitude from zero to maximum without affecting the harmonic elimination results, two identical voltages V1 and V2 have to be combined with the variable symmetrical phase shift [32, 33] (**Figure 13**).

The variable phase-shift symmetry maintains stable phase of the resulting output voltage at the load during the amplitude regulation. This method (also known as outphasing) was originally developed for AM transmitters and assumed two sinusoidal combining voltages, now widely used for high-efficiency communication transmitters [34]. The amplitude of the combined output voltage Vc depends upon the phase shift φ as shown on phasor diagram (**Figure 13**):

$$V_c = 2\,V_0\,Sin\varphi \tag{11}$$

For the high-power applications, the switch-mode converters are used instead of the sine generators (**Figure 14**) where two sets of power modules Mod A and Mod B are controlled from two multiple outputs of fixed delay modules A and B. Delay modules may be shift registers or digital delay lines (for RF transmitters). Inputs of delay lines controlling leading vector A and lagging vector B are connected to the outputs of symmetrical phase-shift modulator with amplitude control voltage and sinusoidal carrier voltage inputs. The shown output transformers are the simplest way to combine output voltages and to form regulated output. For power conversion with the strict requirements to the harmonic content of the regulated sinusoidal voltage at the output, each of two DC/AC converters shall comply with those requirements, and the number of modules is chosen based in **Figure 11** data.

Sinusoidal carrier voltage (not triangular or sawtooth voltage) provides linear modulation characteristic for the resulting fundamental harmonic, which allows direct control of the RF output pulses restored at the resonant antenna without the real-time negative feedback. In some cases the correction of the total system gain depending on the DC bus voltage may be done between RF pulses [29].

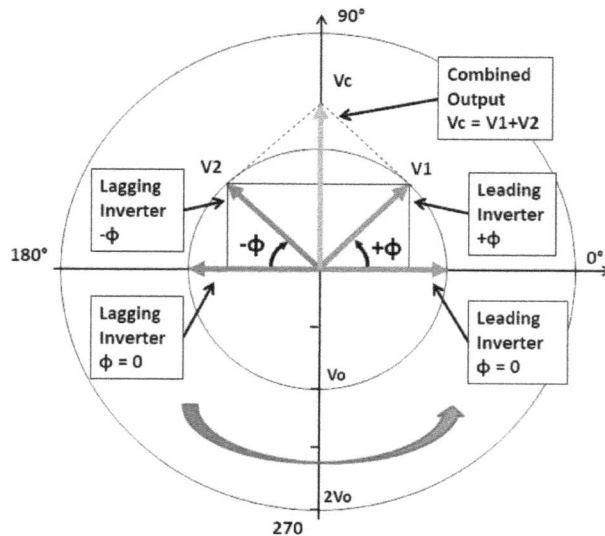

Figure 13. Outphasing modulation phasor diagram for the fundamental harmonic.

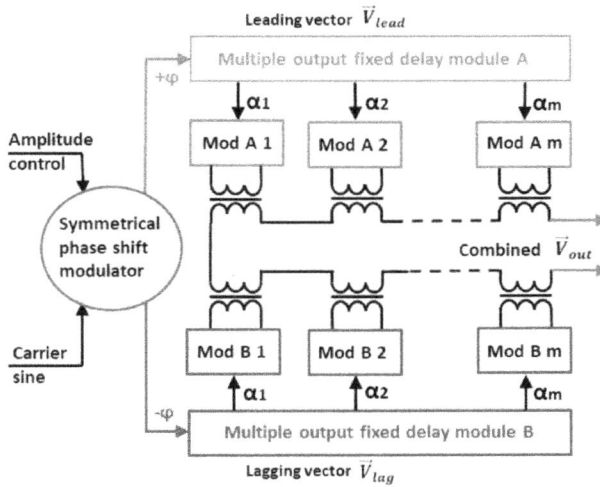

Figure 14. DC/AC multimodule multilevel converter with outphasing amplitude control.

6. RF transmitter current harmonics and life expectancy

The simplified schematic of the two module switch-mode outphasing transmitters is shown in **Figure 15** [35]. Two full-bridge converters, named leading and lagging, have their output voltages combined using their output transformers TX1 and TX2. The load is a resonant antenna consisting of the antenna inductor La, parallel-tuned capacitor Ca and resistor Ra representing antenna losses and defying antenna Q. The load impedance has its maximum at the operating frequency of the parallel resonance equal to Ra. For the higher harmonics of the output voltage, antenna impedance is capacitive and drops at higher frequency. Higher harmonics of the transmitter output current are limited by the output filters Lf and Cf. This filter is tuned to the series resonance at the operating frequency to introduce minimum output filter voltage drop which is proportional to the output current.

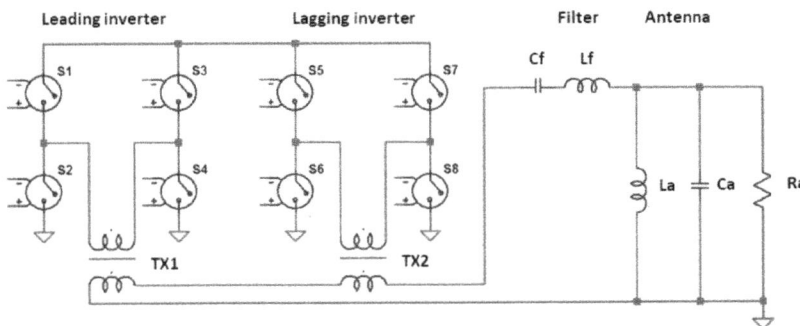

Figure 15. Switch-mode outphasing transmitter.

The transferring function transmitter output voltage to the antenna voltage is shown in **Figure 16**. The "saddle characteristic" has two poles reflecting two resonances at lower and higher frequencies than the operating frequency. The amplitude of those peaks depends upon the antenna and filter losses as shown for Q = 10 and Q = 100. The box at the operational frequency 0.5 MHz sets the acceptable system bandwidth which is necessary for correct restoring of the envelope of RF pulses used in NMR tools for the proton excitation. For RF pulses with Hann envelope with duration of 25us minimum bandwidth to correctly recover amplified antenna pulses, the bandwidth of the system output filter plus antenna should be 40 kHz minimum; with reliable margin for wide temperature range, it should be 80 kHz with flatness 5% or 0.5db (**Figure 17**).

Increase of the system bandwidth benefits the metrological parameters of the tool and accuracy in the replication of the shortest RF pulses, used for the hydrogen nuclei excitation, and makes possible to use the same filter for more than one frequency without the filter capacitor switched. But there is another drawback: the higher-frequency resonance (the right pole) amplifies higher harmonics of the output current due to the drop of the filter/antenna input impedance (**Figure 18**). The biggest harmonic in the 50% duty cycle module output voltage is the third harmonic, and the third harmonic of the output current may significantly exceed the amplitude of the fundamental harmonic. The smaller is output filter inductance the higher are values of the unwanted higher harmonics in the output current and the higher are conductive losses resulting in the component temperature rise. The output filter design is a compromise between the bandwidth and accuracy vs. losses and reliability. Higher harmonic elimination removes the cause of the excessive output current and significantly reduces power losses for the same number of switches [28].

To assess the advantage of the discussed harmonic elimination method, three 500 kHz transmitters operating identical loads (2uH, 52 nF output filter and 1uH, 100 nF, 90ohm, Q 30 antenna) and using the same number of power MOSFET switches (32 total) but different topologies and control signals corresponding to Case A, Case B and Case C configurations (**Figure 10**) were simulated for the standard for NMR pulse train with 25us and 50us Hann envelope, 1000Vmax and 20% duty cycle (**Figure 19**) [29].

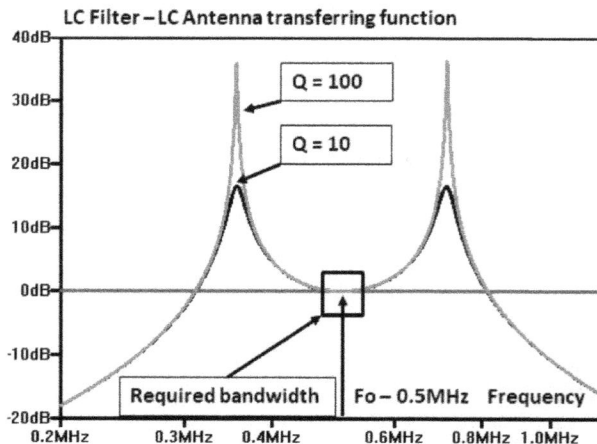

Figure 16. Transferring function filter plus antenna.

Output voltage

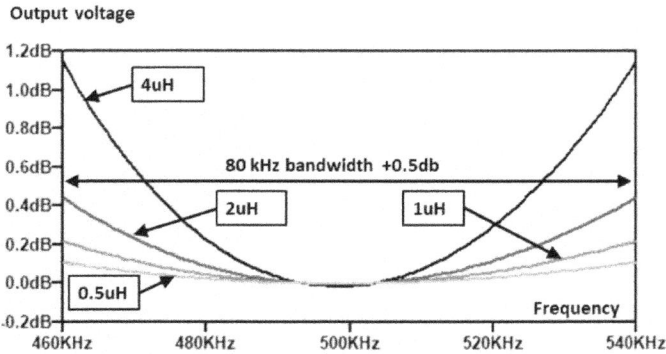

Figure 17. Bandwidth vs. filter inductance.

Standard NMR excitation pulse sequence consists of the multiple identical pulses with the equal period and the first pulse with half duration and the same amplitude Vp =1000 V as shown on **Figure 19**.

The harmonics of the transmitter output current depend upon the voltage harmonics in the transmitter output signal. **Figure 20** shows the values of the first seven most important harmonics of the output current for transmitters based on the simplest two module Case A configurations [35], four module-enhanced transmitter Case B with eliminated the 3rd harmonic [50] and eight module transmitter Case C with eliminated the 3rd and 5th harmonics. Proper waveforms of the leading or lagging voltages marked as A, B and C are provided in **Figure 22**. Elimination of higher harmonics decreases the total output RMS current I_{out} per switch and resulting conductive power dissipation P_s:

$$P_s = P_{sw} + I_{out}^2 * R_{dson} \tag{12}$$

Current

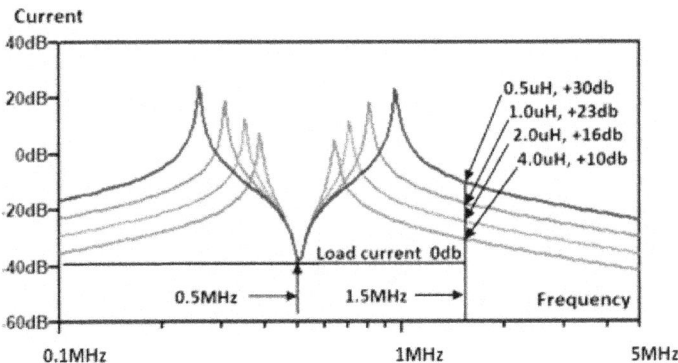

Figure 18. Transmitter output current of the third harmonic compared to the first one vs. filter inductance.

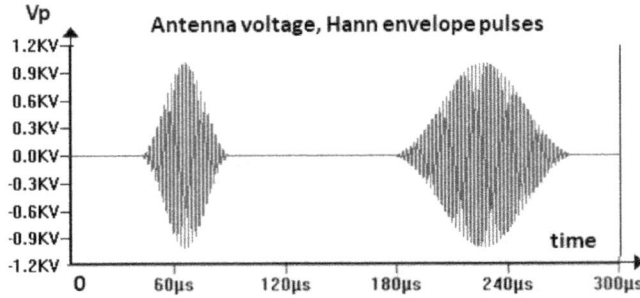

Figure 19. Antenna voltage (Vp(t)) and RF pulses with Hann envelope.

where P_s is the power switch total power dissipation, P_{sw} is the power of switching losses, I_{rms} is the RMS value of switch current and R_{dson} is the switch resistance drain to source, which for silicon MOSFET is the function of the temperature and current [36].

Additional temperature rise ΔT_j of the switch die over the ambient temperature T_A depends on the thermal resistance R_{th} from the switch die to the ambient temperature T_A and current I_{out}:

$$T_j = T_A + \Delta T = T_A + R_{th} * \left(P_{sw} + I_{out}^2\right) \tag{13}$$

The die temperature T_j affects the switch life expectancy t_v according to the Arrhenius law:

$$t_v = C * e^{\frac{E_A}{kT_j}} \tag{14}$$

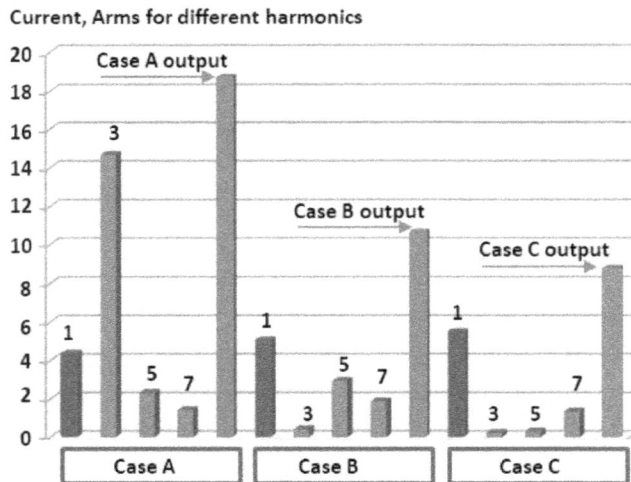

Figure 20. Higher harmonic (blue) elimination decreases total output current (red).

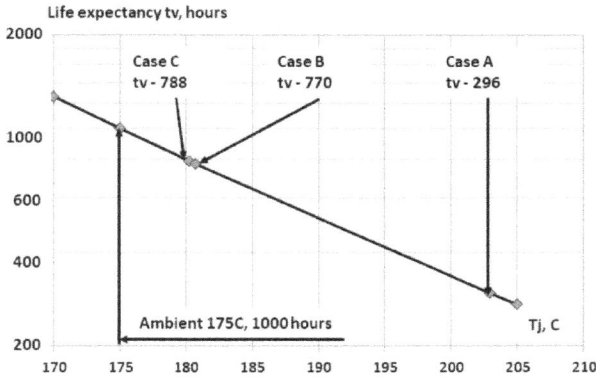

Figure 21. Life expectancy tv vs. transmitter topology.

where t_v is the switch life expectancy in time units, years or hours; T is the absolute temperature; E_A is the apparent activation energy in general depending on T, recommended value 1.8 eV [37]; C is a constant and k is the Boltzmann constant [38].

Additional temperature rise ΔT_j decreases life expectancy at T_j by the accelerator factor (AF) compared to the life expectancy at T_A:

$$AF = \frac{t_{v(T_j)}}{t_{v(T_A)}} = e^{\frac{E_A}{k}\left(\frac{1}{T_j} - \frac{1}{T_A}\right)} \tag{15}$$

Implementation of the sequential harmonic elimination decreases output current and related temperature rise, which results into the improvement of the transmitter reliability in terms of the life expectancy. The life expectancy (shown in **Figure 21**) changes from 1000 h at 175°C, taken as a reference point, down to 296 h for Case A, to 770 h for Case B and to 788 h for Case C. All transmitters use 32 power MOSFET switches; the difference is in the topology, control method and number of the additional RF output transformers, which easily fit the pressure housing and do not generate significant amount of heat.

Switching from classic transmitter topology Case A to Case B increases life expectancy 2.6 times, from Case A to Case C—2.66 times. It is clear that Case B transmitter is the best solution in terms of reliability improvement vs. extra cost and complexity. The benefits include not only increases of the life expectancy but also increases of tool accuracy, improving EMI conditions in the confined space of the power train placed in the high-pressure housing due to the spread in time switch operation [29].

7. Conclusion

The sequential harmonic elimination provides a simple method of removing higher harmonics from the output voltage of the multimodule multilevel converters operating with the fundamental switching frequency and using identical modules, generating 50% duty cycle output voltages. A simple algorithm for the control circuitries used to eliminate harmonics and regulate output voltage from zero to maximum maintaining stable phase is discussed. A simple expression for THD of the output voltage vs. the number of eliminated harmonics, derived from simulation results, is provided for design evaluation against IEEE 519 standard requirements. The application of this method to the NMR transmitters operating in the high-temperature environment eliminates the most dangerous output current harmonics and shows more than twice the gain in the life expectancy. This method was validated for NMR downhole logging equipment, and two patents were granted.

Author details

Alexey Tyshko

Address all correspondence to: alex.tyshko@gmail.com

Baker Hughes, a GE Company, Global Research Center, Oklahoma City, OK, USA

References

[1] Patel HS, Hoft RG. Generalized techniques of harmonic elimination and voltage control in thyristor inverters: Part I harmonic elimination. IEEE Transactions on Industry Applications 9, May 1973;**IA**:310-317. DOI: 10.1109/TIA.1973.349908

[2] Baker, R, Bannister L. Electric power converter. Patent US 3,867,643 A USA. 1974. https:/ www.google.com/patents/US3867643

[3] Bhagwat P, Stefanovic V. Generalized structure of a multilevel PWM converter. IEEE Transactions on Industry Applications. 1983;**19**(6):1067-1069. DOI: 10.1109/TIA.1983.4504335

[4] Carrara G, Gardella S, Marchesoni M, Salutari R, Scuitto G. A new multilevel PWM method: A theoretical analysis. IEEE Transactions on Power Electronics. July 1992;**7**:497-505. DOI: 10.1109/63.145137

[5] Menzies RW, Zhuang Y. Advanced static compensation using a multilevel GTO thyristor inverter. IEEE Transactions on Power Delivery. April 1995;**10**(2):732-738. DOI: 10.1109/61.400858

[6] Cengelci E, Enjeti P, Singh C, Blaabjerg F, Pederson J. New medium voltage PWM inverter topologies for adjustible speed AC motor drive systems. Proceedings IEEE APEC' 98 1998;**2**:565-571. DOI: 10.1109/APEC.1998.653955

[7] Lai J-S, Peng F. Multilevel converters: A new breed of power converters. IEEE Transactions on Industry Applications. May/June 1996;**32**(3):509-517. DOI: 10.1109/28.502161

[8] Li L, Czarkowski D, Liu Y, Pillay P. Multilevel selective harmonic elimination PWM technique in series-connected voltage inverters. IEEE Transactions on Industry Applications Jan/Feb 2000;**36**(1):160-170. DOI: 10.1109/28.821811

[9] Rodriguez J, Lai J-S, Peng FZ. Multilevel inverters: A survey of topologies, controls, and applications. IEEE Transactions on Industrial Electronics. August 2002;49(4):724-738. DOI: 10.1109/TIE.2002.801052

[10] Tolbert L, Peng F, Habetler T. Multilevel converters for large electric drives. IEEE Transactions on Industrial Applications. Jan/Feb 1999;35:36-44. DOI: 10.1109/28.740843

[11] Du Z, Tolbert L, Chiasson J. Harmonic Elimination for Multilevel Converter with Programmed PWM Method. IEEE IAS; 2004. pp. 2210-2215. DOI: 10.1109/IAS.2004.1348783

[12] Rodrigues J, Bernet S, Wu B, Pontt J, Kouro S. Multilevel voltage-source-converter topologies for industrial medium-voltage drives. IEEE Transactions on Industrial Electronics. December 2007;54:2930-2945. DOI: 10.1109/TIE.2007.907044

[13] Rodriuez P, Bellar MD, Munoz-Aguilar RS, Busgues-Monge S, Blaabjerg F. Multilevel-Clamped Multilevel Converters (MLC). 3, s.l. IEEE. 2012;27:1055-1060. DOI: 10.1109/TPEL.2011.2172224

[14] Dahidah M, Agelidis V. Selective harmonic elimination PWM control for cascaded multilevel voltage source converters: A generalized formula. IEEE Transactions on Power Electronics. 2008;23(4):1620-1630. DOI: 10.1109/TPEL.2008.925179

[15] Dahidah M, Konstantinou G, Agelidis VA. Review of multilevel selective harmonic elimination PWM: Formulations, solving algorithms, implementation and applications. IEEE Transactions on Power Electronics. 2014;99:1-16. DOI: 10.1109/TPEL.2014.2355226

[16] Darus R, Konstantinou G, Pou J, Ceballos S, Agelidis V. Comparison of Phase-Shifted and Level-Shifted PWM in the Modular Multilevel Converter. In: IEEE International Power Electronics Conference (IPEC–ECCE Asia) 2014. Hiroshima, Japan: IEEE; 2014. DOI: 10.1109/IPEC.2014.6870039

[17] Gnanasambandam K, Rathor AK, Edpuganti A, Srinivasan D, Rodriguez J. Current-Fed Multilevel Converters: An Overview of Circuit Topologies, Modulation Techniques, and Applications. IEEE Transactions on Power Electronics. 2017;32:3382-3402. DOI: 10.1109/TPEL.2016.2585576

[18] Ilves K, Antonopoulos A, Norrga S, Nee HP. A new modulation method for the modular multilevel converter allowing fundamental switching frequency. Proc. IEEE ICPE (ECCE Asia). June 2011:991-998. DOI: 10.1109/ICPE.2011.5944672

[19] Ilves K, Harnefors L, Norrga S, Nee H-P. Analysis and operation of modular multilevel converters with phase-shifted carrier PWM. IEEE Transactions on Power Electronics. Jan 2015;30(1):268-296. DOI: 10.1109/TPEL.2014.2321049

[20] Konstantinou G, Pulikanti S, Agelidis V. Harmonic Elimination Control of a Five-Level DC-AC Cascaded H-Bridge Hybrid Inverter. s.l. 2nd IEEE International Symposium on Power Electronics for Distributed Generation Systems, 2010. 2010. DOI: 10.1109/PEDG.2010.5545762

[21] Li B, Yang R, Xu D, Wang G, Wang W, Xu D. Analysis of the phase-shifted carrier modulation for modular multilevel converters. IEEE Transactions on Power Electronics. Jan 2015;30(1):297-310. DOI: 10.1109/TPEL.2014.2299802

[22] Chiasson JN, Tolbert LM, Mckenzie KJ, Du Z. Elimination of harmonics in a multilevel converter using the theory of symmetric polynomials and resultants. IEEE Transactions on Control Systems Technology. 2005:216-223. DOI: 10.1109/CDC.2003.1271691

[23] IEEE Standard 519-2014. IEEE Recommended Practices and Requirements for Harmonic Control in Electrical Power Systems, New York. https://edisciplinas.usp.br/pluginfile. php/2587631/mod_resource/content/2/IEE%20Std%20519-2014.pdf

[24] Capua CD, Romeo E. A smart THD meter performing an original uncertainty evaluation procedure. IEEE Transactions on Instrumentation and Measurement. Aug. 2007;56(4):1257-1264. DOI: 10.1109/TIM.2007.899895

[25] Blagouchine I, Moreau E. Analytic method for the computation of the total harmonic distortion by the Cauchy method of residues. IEEE Transactions on Communications. 2011;59(9):2478-2491. DOI: 10.1109/TCOMM.2011.061511.100749

[26] Tyshko A. Multi-Vector Outphasing DC to AC Converter and method. US Patent 9680396. granted June 13, 2017. https://www.google.com/patents/US9680396

[27] Tyshko A. Multi-Vector Outphasing Provides High Power, Low Harmonics. IEEE Digital Library ELNANO-2014. Kiev, Ukraine: IEEE, 2014. Electronics and Nanotechnology (ELNANO), 2014 IEEE 34th International Conference. pp. 416-420. DOI: 10.1109/ ELNANO.2014.6873437

[28] Tyshko A. Excitation of the Resonant Loads with the Multi-Vector Synthesized Sinusoidal Voltage Decreases Conduction Losses and Improves Reliability. Vol. 19. Kiev, Ukraine: Electronics and Communications, Power Electronics Section; 2014. pp. 37-42 ISSN 1811-4512

[29] Tyshko A, Balevicius S, Padmanaban S. An increase of a down-hole nuclear magnetic resonance Tool's reliability and accuracy by the cancellation of a multi-module DC/AC Converter's Output's higher harmonics. IEEE Access. 2016;4:7912-7920. DOI: 10.1109/ ACCESS.2016.2624498

[30] Tyshko A. Enhanced Transmitter and Method for a Nuclear Magnetic Resonance Logging Tool. Patent US 9,405,035. Jan 10, 2012. Patent granted Aug 2, 2016; https://www.google. com/patents/US9405035

[31] Tyshko A. DC to AC 3 Phase Modular Multilevel Conversion Using Chireix Outphasing Method. Kiev, Ukraine: IEEE, 2015. Electronics and Nanotechnology (ELNANO), 2015 IEEE 35th International Conference. pp. 539-542. DOI: 10.1109/ELNANO.2015.7146946

[32] Chireix H. High power outphasing modulation. Proceedings of IRE. 1935;23:11

[33] Doherty WH. A new high efficiency power amplifier for modulated waves. Proceedings of IRE. 1936;24:9

[34] Perreault DJ. A new power combining and Outphasing modulation system for high efficiency power amplification. IEEE Transactions on Circuits and Systems. Aug 2011;58(8):1713-1726 http://dx.doi.org/10.1109/TCSI.2011.2106230

[35] Dent, Paul Wilkinson. Hybrid Chireix/Doherty amplifiers and methods. Patent US 6,133,788 USA, Apr 2, 1998. https://www.google.com/patents/US6133788

[36] Tyshko A. Analyze MOSFET parameter shifts near maximum operating temperature. Electronic Design. March 2014;**17**:62-65 EDN (ISSN 0012-7515)

[37] Bayle F, Metta A. Temperature acceleration models in reliability predictions, Justification and Improvements. In: Proc. Reliability and Maintainability Symp. San Jose, CA, USA. Jan. 2010. pp. 1-7 http://www.reliasoft.com/pubs/2010_RAMS_temperature_accel-eration_models.pdf

[38] Calixto E. Gas and Oil Reliability Engineering: Modeling and Analysis. Oxford, UK: Gulf Professional Publishing; 2012 ISBN: 9780128054277

Chapter 3

Harmonics Temporal Profile in High-Voltage Networks: Case Study

Mohammed H. Albadi, Rashid S. Al Abri,
Amer S. Al Hinai and Abdullah H. Al-Badi

Additional information is available at the end of the chapter

http://dx.doi.org/10.5772/intechopen.72568

Abstract

This chapter presents a case study about harmonics measurements in high-voltage networks. Measurements were conducted at two locations in the main interconnected system (MIS) of Oman. Voltage and current THDs were recorded for a period of 1 week. The power quality analyzer was set to record required data for a period of 1 week, and the observation period for each recorded value is 10 minutes. At the first location, the grid station (132/33) is feeding industrial as well as other customers. The second grid station (220/132/33 kV) is dedicated to large industrial customers including arc furnaces and rolling mills. The power quality analyzer was installed at the 132 kV side of power transformers at both locations. Recorded data are analyzed, and temporal harmonics profiles are studied. A clear temporal variation of harmonics similar to that of aggregate load and local voltage profiles was observed at the grid station feeding mixed residential and industrial loads. However, this correlation between system load and harmonics profile diminishes at the grid station dedicated for heavy industrial loads.

Keywords: harmonics, measurements, load profile, temporal profile, standards

1. Introduction

Harmonics are caused by non-linear loads, which draw a non-sinusoidal current from a sinusoidal voltage source. Examples of such harmonic-producing loads, which are used extensively in the industry, include inverters, DC converters, electric arc furnaces, static VAR compensators, switch-mode power supplies (SMPS), and AC or DC motor drives. Other loads such as photocopiers, personal computers, laser printers, fax machines, battery chargers, fluorescent lamps, and UPSs are also a source of harmonics that can be found in the commercial sector [1, 2].

ntechOpen

Large industrial loads are often connected to transmission networks due to large power requirements. Such loads are often non-linear and may include rolling mills driven by variable speed drives or could be an arc furnace. These non-linear loads are sources of harmonics. Harmonics that propagates from the industrial loads degrades the power quality at the electrical system. Harmonics can cause problems to different electrical equipment such as generators, motors, transformers, capacitors, and cables. In addition, it can lead to reduced capacity and efficiency of power systems.

The harmonic content in any network varies with time depending on the share of non-linear loads as well as system status. Examining temporal harmonic profile can help understanding system performance at different loading conditions. To examine the temporal profile, harmonics measurements were conducted at two industrial load grid stations of Oman Electricity Transmission Company (OETC) network in the Main Interconnected System (MIS) of Oman.

2. Harmonic limits according to the international standards and Oman's national regulations

In this section, the indices conventionally used for measurement of voltage and current harmonic distortion and the harmonic distortion limits placed in IEEE standard 519, IEC standard 61000-3-6 and Oman's national regulations are presented.

2.1. Voltage and current harmonic distortion indices

The total harmonic distortion (THD) is used to define the effect of harmonics on the power system voltage. IEEE 519-2014 defines the THD as "the ratio of the root mean square of the harmonic content, considering harmonic components up to the 50th order and specifically excluding interharmonics, expressed as a percent of the fundamental". In other words, the THD is the contribution of all harmonics to the fundamental. The THD is calculated as described by the following formula:

$$THD = \frac{\sqrt{\sum_{h=2}^{h_{max}} M_h^2}}{M_1}$$ (1)

where M_1 is the rms value fundamental component of the voltage or current signal.

To evaluate the current harmonic distortion, the total demand distortion (TDD) is commonly used. IEEE 519-2014 defines the TDD as "the ratio of the root mean square of the harmonic content, considering harmonic components up to the 50th order and specifically excluding interharmonics, expressed as a percent of the maximum demand current"

$$TDD = \frac{\sqrt{\sum_{h=2}^{h_{max}} I_h^2}}{I_m} = \frac{\sqrt{I^2 - I_1^2}}{I_m}$$ (2)

where I_m is the maximum demand load current and I_1 is the rms value of the fundamental component.

2.2. IEEE standard 519 harmonic distortion limits

IEEE standard 519 gives harmonic distortion limits for both the current and the voltage signals in power systems [3]. **Tables 1** and **2** show relevant voltage and current distortion limits. The allowable voltage THD is based on the voltage level, while the current TDD limit is given based on the voltage level and the ratio of the short circuit current to the rated load current.

According to IEEE STD 519-2014, statistical analysis of 1-week short-time harmonic measurements is required to calculate the 95th and 99th percentile values for comparison with the recommended limits. While current harmonics are evaluated based on 95th and 99th percentiles, voltage harmonics are evaluated based on 95th percentile only.

2.3. IEC standard 61000-3-6 harmonic distortion limits

IEC Standard 61000-3-6 specifies the allowable harmonic distortion limits as shown in **Table 3** [4].

PCC voltage	Individual harmonic magnitude (%)	THD (%)
V ≤ 1 kV	5	8
1 < V ≤ 69 kV	3	5
69 < V ≤ 161 kV	1.5	2.5
V > 161 kV	1	1.5

Table 1. Voltage THD limits according to IEEE 519-2014 [3].

Voltage	Isc/I(load)	TDD (%)	<11	11 ≤ h < 17	17 ≤ h < 35	35 ≤ h ≤ 50
≤69 kV	<20	5	4	2	1.5	0.6
	20–50	8	7	3.5	2.5	1
	50–100	12	10	4.5	4	1.5
	100–1000	15	12	5.5	5	2
	>1000	20	15	7	6	2.5
69–161 kV	<20	2.5	2	1	0.75	0.3
	20–50	4	3.5	1.75	1.25	0.5
	50–100	6	5	2.25	2	0.75
	100–1000	7.5	6	2.75	2.5	1
	>1000	10	7.5	3.5	3	1.25
>161 kV	<25	1.5	1	0.5	0.38	0.1
	25–50	2.5	2	1	0.75	0.3
	≥50	3.75	3	1.5	1.15	0.45

Table 2. Current distortion limits according to IEEE 519-2014 [3].

Odd harmonics non-multiple of three (%)			Odd harmonics multiple of three (%)			Even harmonics (%)		
h	MV	HV-EHV	H	MV	HV-EHV	h	MV	HV-EHV
5	5	2	3	4	2	2	1.8	1.4
7	4	2	9	1.2	1	4	1	0.8
11	3	1.5	15	0.3	0.3	6	0.5	0.4
13	2.5	1.5	21	0.2	0.2	8	0.5	0.4
$17 \leq h \leq 49$	$1.9\frac{17}{h} - 0.2$	$1.2\frac{17}{h}$	$21 < h \leq 45$	0.2	0.2	$10 \leq h \leq 50$	$0.25\frac{10}{h} + 0.22$	$0.19\frac{10}{h} + 0.16$

Table 3. IEC 61000-3-6 voltage harmonic limits [4].

PCC voltage	Individual harmonic magnitude (%)	VTHD (%)
Low voltage (415 V)	—	2.5
Distribution level (11, 33, 66 kV)	1.5	2
Transmission level (132, 220 kV)	1.5	2

Table 4. Voltage THD limits according to national Omani codes.

2.4. Harmonic distortion limits according to Oman's national regulations

Allowable harmonic distortion levels in Oman are dictated by the grid code [5] and the distribution code [6] for high-voltage and medium-voltage networks, respectively. The grid code specifies that the maximum THD should not exceed 2% with no individual harmonic greater than 1.5% for transmission networks (220 and 132 kV). The distribution code dictates that the maximum THD in distribution networks (66, 33 and 11 kV) systems should not exceed 2.0% with no individual harmonic greater than 1.5%. For low voltage line (415 V), the total harmonic distortion limit is 2.5%. Individual harmonic distortion level should be below 1.5% for both transmission and distribution networks (**Table 4**).

3. Harmonics temporal profile at grid station A

3.1. Harmonics measurements at grid station A

Figure 1 shows MIS system and the locations of grid stations under study. The grid station A consists of four 132/33 kV parallel transformers (TX1 to TX4). Each transformer is 75 MVA capacity. As the grid station is located adjacent to a 600 MW power generation facility, the maximum short circuit level on 132 kV side is 27.54 kA. The grid station is supplying electricity to a cement factory, an industrial area, industrial area housing, a university campus, and a hospital.

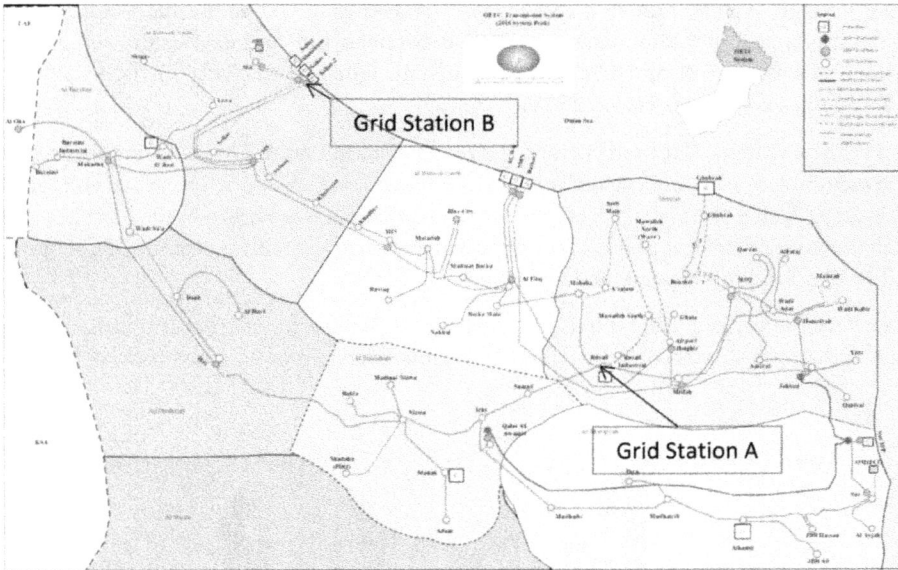

Figure 1. Main interconnected system [7].

Measurements were conducted using Hioki 3196 Power Quality Analyzer Meter [8]. The current clamps were installed on transformer one (TX1). Ten-minute average values were recoded over a period of 1 week starting from 17th of January 2012. This study was part of power quality study in MIS [9–12]. A summary of measurements is presented in **Figure 2**.

TIME PLOT · RMS U-THD, CH1			TIME PLOT · RMS I-THD, CH1		
	AVE			AVE	
A 01/24 11:41:00		0.83	A 01/24 11:41:00		1.20
B 01/17 11:51:00		0.79	B 01/17 11:51:00		1.18
· 7 00:00:00		· 0.04	· 7 00:00:00		· 0.01
MAX values		1.10	MAX values		2.31
AVE values		0.83	AVE values		1.41
MIN values		0.52	MIN values		0.84
TIME PLOT · RMS U-THD, CH2			TIME PLOT · RMS I-THD, CH2		
	AVE			AVE	
A 01/24 11:41:00		0.70	A 01/24 11:41:00		1.23
B 01/17 11:51:00		0.64	B 01/17 11:51:00		1.30
· 7 00:00:00		· 0.06	· 7 00:00:00		0.07
MAX values		0.92	MAX values		2.27
AVE values		0.68	AVE values		1.45
MIN values		0.40	MIN values		0.84
TIME PLOT · RMS U-THD, CH3			TIME PLOT · RMS I-THD, CH3		
	AVE			AVE	
A 01/24 11:41:00		0.75	A 01/24 11:41:00		1.53
B 01/17 11:51:00		0.72	B 01/17 11:51:00		1.46
· 7 00:00:00		· 0.03	· 7 00:00:00		· 0.07
MAX values		1.07	MAX values		2.51
AVE values		0.80	AVE values		1.62
MIN values		0.49	MIN values		1.03

a)   b)

Figure 2. Summary of THD measurements at grid station a. (a) Voltage THD (b) current THD.

For the voltage signals, **Figure 1** shows that the average THD at the high-voltage side is between 0.84 and 0.68%, the maximum THD is between 1.07 and 0.92%, and the minimum THD level is between 0.40 and 0.52%. For the current signals, the THD at the high-voltage side (132 kV) ranges between 0.84 and 2.52%.

Histograms of voltage THD and current TDD at grid station A are presented in **Figure 3**. Since the measurements were conducted at 132 kV voltage level, the corresponding voltage THD limit is 2.5%. Using the voltage THD histograms presented in **Figure 3**, the 95th percentiles are calculated for different phases. A comparison between the 95th percentile of voltage THD and IEEE Std 519 limit is presented in **Table 5**.

To calculate the current TDD, Eq. (2) is used. The TDD is a function of individual harmonics and maximum demand load current. Using measurements, the maximum demand load

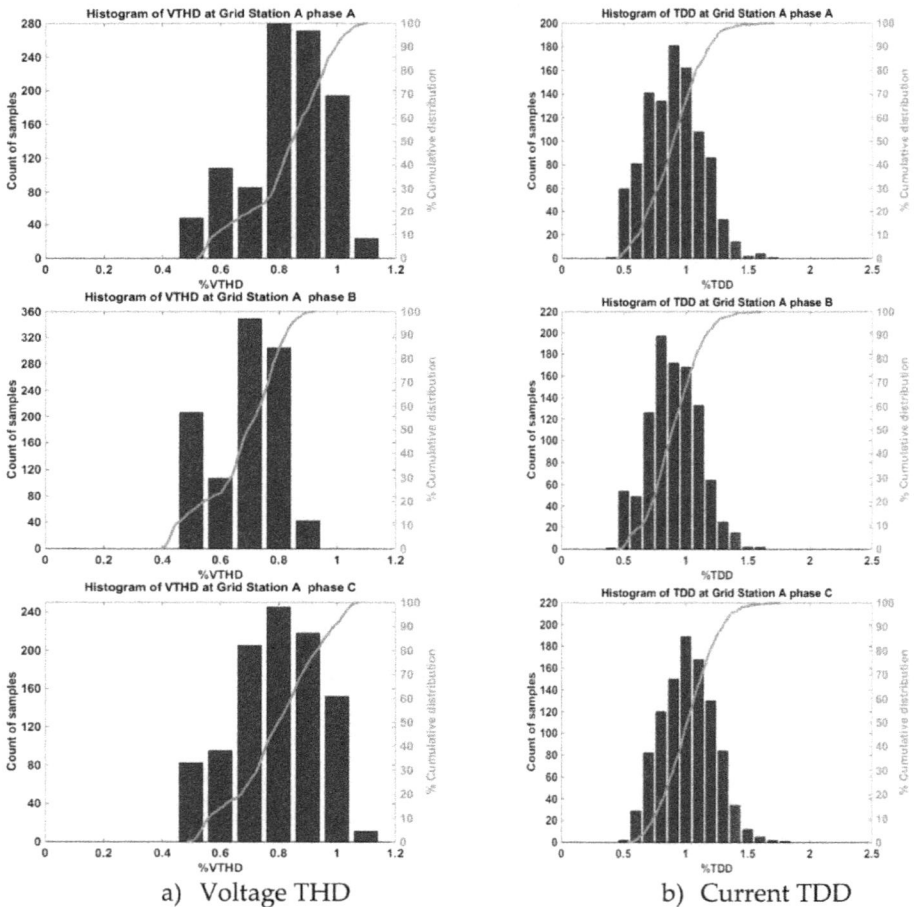

a) Voltage THD   b) Current TDD

Figure 3. Histograms of THD measurements at grid station A. (a) Voltage THD (b) current TDD.

Phase	VTHD P95	VTHD IEEE limits
A	1.03	2.5
B	0.85	2.5
C	1.03	2.5

Table 5. Measured voltage THD 95th percentile versus IEEE Std 519-2014.

current is 369 A. The short circuit level at grid station A is obtained from OETC capability statement. The TDD limit based on I_{SC}/I_L = 39 is 4%. A comparison between the 95th and 99th percentiles of current TDD and IEEE Std 519 limits is presented in **Table 6**.

3.2. Harmonics temporal voltage profile at grid station A

When system loading increases, more voltage drop occurs. Therefore, the temporal voltage profile reflects the daily load profile. The voltage profile measured at grid station A is presented in **Figure 4**.

The voltage THD exhibits a daily profile similar to that of the load as demonstrated in **Figure 5**. Considering individual harmonics, the figure shows that the dominant harmonic components are the 5th and the 7th followed by the 3rd and 11th. There is a small trace of the 13th and 9th harmonics.

3.3. Current harmonics temporal profile at grid station A

Unlike voltage THD, current THD does not exhibit a clear daily profile. **Figure 6** demonstrates that the dominant harmonic components are the 5th and the 3rd. There is a small trace of the 7th harmonics. It is worth mentioning that a cement factory is connected to this feeder.

3.4. Reasons for temporal variations

The temporal variations of the voltage and current THD are associated with the variations of the harmonic-producing loads. Normally, the individual harmonic distortion is linked to specific harmonic-producing loads. Station A is a grid station that is connected to a cement factory and an industrial area where many non-linear loads are fed from this station. Such loads are variable frequency drive (VFD) and switch-mode power supplies (SMPS), in which both are considered main sources of the 5th harmonics (H5). This explains the dominance of the 5th and the 3rd harmonic currents contamination.

Phase	TDD P95	TDD P99	TDD IEE limits
A	1.25	1.42	4
B	1.24	1.30	4
C	1.36	1.52	4

Table 6. Measured current TDD 95th and 99th percentiles vs. IEEE Std 519-2014.

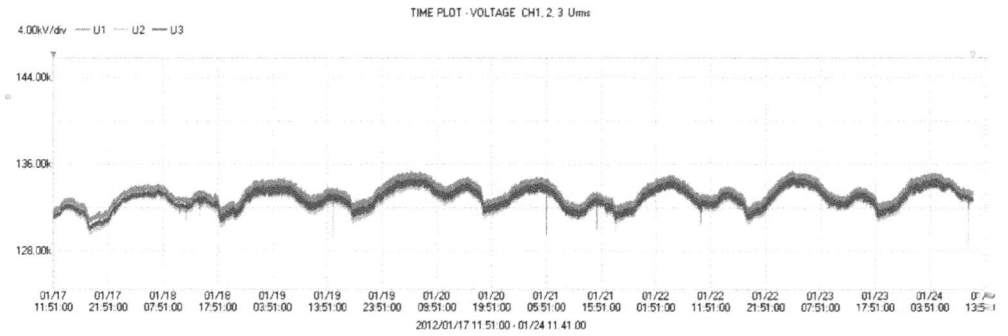

Figure 4. Voltage temporal profile at grid station A.

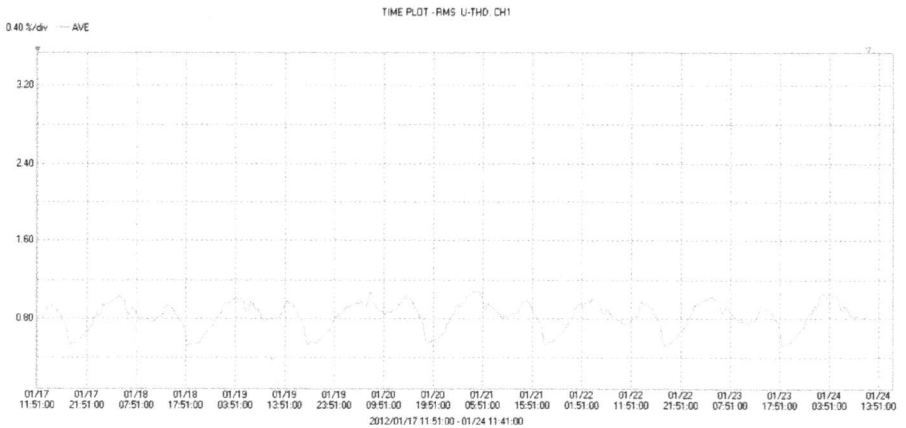

(a)

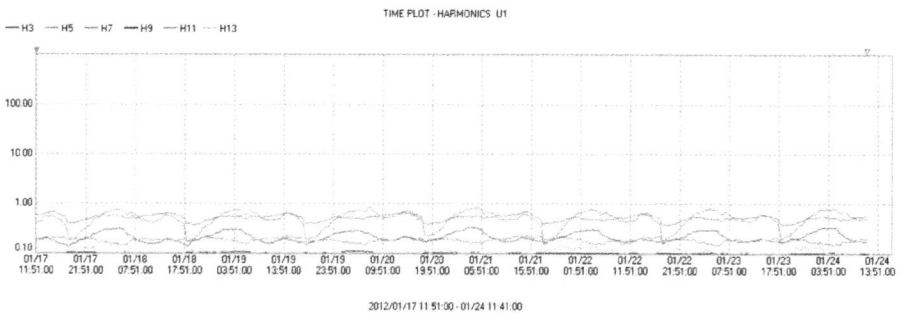

(b)

Figure 5. Voltage harmonics temporal profile at grid station A. (a) Voltage THD and (b) individual harmonics.

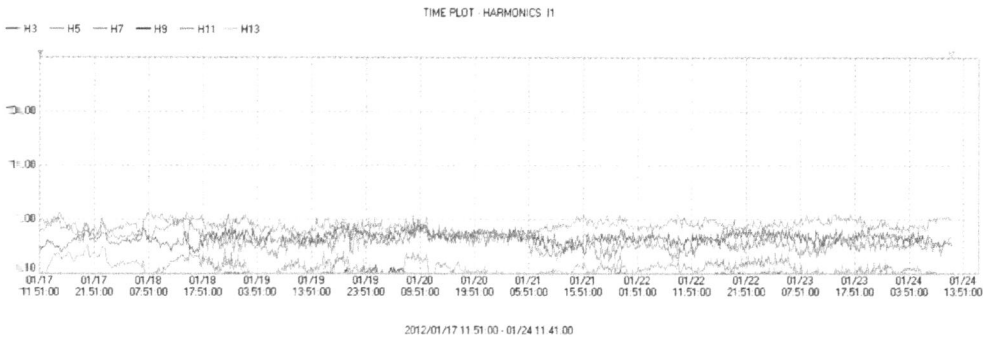

Figure 6. Current harmonics temporal profile at grid station A.

4. Harmonics temporal profile at grid station B

4.1. Harmonic measurements at grid station B

Grid station B consists of two 220/132 kV parallel transformers (TX1 and TX2). Each transformer is 500 MVA capacity. The grid station B is located close to two power generation facilities. The maximum short circuit level on 132 kV side at the time of measurements is 17.96 kA. The grid station is supplying electricity to large industrial customers via 132 kV feeders as well as smaller industrial customers via two 132/33 kV transformers. Ten-minute average values were recoded using Hioki 3196 Power Quality Analyzer Meter [8]. The current clamps of the analyzer were connected to a 132 kV feeder supplying a steel smelter. Measurements were recoded over a period of 1 week starting from 28 January 2012. A summary of measurements is presented in Figure 7.

For the voltage signals, **Figure 7** shows that the average THD at the high-voltage side is between 0.97 and 0.90%, the maximum THD is between 0.97 and 0.90%, and the minimum THD level is between 0.28 and 0.32%. For the current signals, the THD at the high-voltage side (132 kV) ranges between 33.09 and 2.28%. Despite the high-current distortion, the voltage THD is small due to the high short circuit capacity.

The current waveform at grid station B is highly distorted as shown in **Figures 7–9**.

It is worth noting that the dominant harmonic components are the 5th, the 7th, the 11th, the 13th and the 17th orders. In addition, there is a clear content of the 2nd, the 3rd, and other harmonic orders.

Histograms of voltage THD and current TDD at station B are presented in **Figure 10**. Using the voltage THD histograms, the 95th percentiles are calculated for different phases. A comparison between the 95th percentile of voltage THD and IEEE Std 519 limit is presented in **Table 7**.

TIME PLOT - RMS U-THD, CH1 AVE		TIME PLOT - RMS I-THD, CH1 AVE	
A 02/04 13:04:00	0.48	A 02/04 13:04:00	9.39
B 01/28 13:14:00	0.46	B 01/28 13:14:00	6.68
- 7 00:00:00	- 0.02	- 7 00:00:00	- 2.71
MAX values	0.97	MAX values	31.37
AVE values	0.43	AVE values	9.96
MIN values	0.28	MIN values	2.28

TIME PLOT - RMS U-THD, CH2 AVE		TIME PLOT - RMS I-THD, CH2 AVE	
A 02/04 13:04:00	0.49	A 02/04 13:04:00	9.42
B 01/28 13:14:00	0.50	B 01/28 13:14:00	6.62
- 7 00:00:00	0.01	- 7 00:00:00	- 2.80
MAX values	0.96	MAX values	33.09
AVE values	0.46	AVE values	10.06
MIN values	0.31	MIN values	2.30

TIME PLOT - RMS U-THD, CH3 AVE		TIME PLOT - RMS I-THD, CH3 AVE	
A 02/04 13:04:00	0.51	A 02/04 13:04:00	9.01
B 01/28 13:14:00	0.44	B 01/28 13:14:00	6.37
- 7 00:00:00	- 0.08	- 7 00:00:00	- 2.64
MAX values	0.90	MAX values	29.54
AVE values	0.45	AVE values	9.57
MIN values	0.32	MIN values	2.28

a)   b)

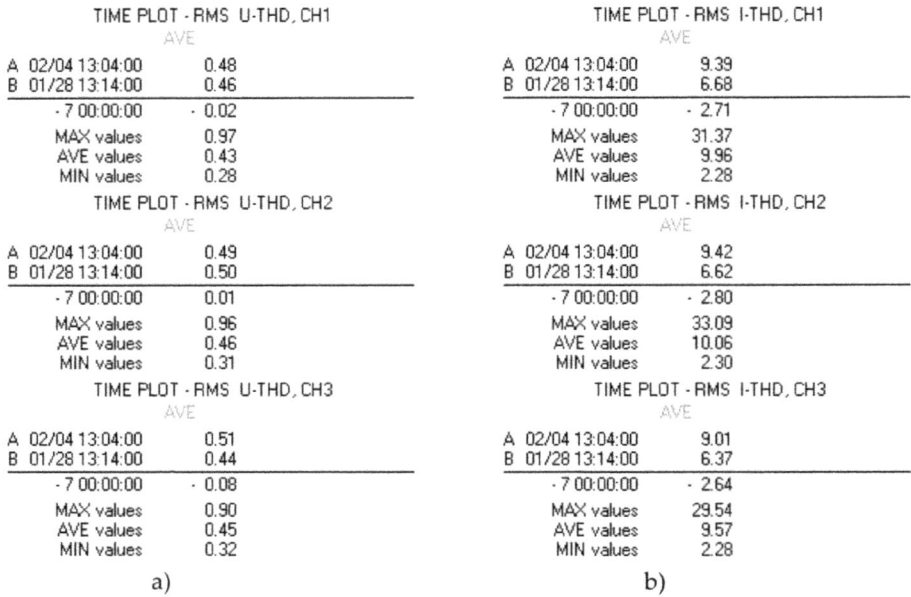

Figure 7. Summary of THD measurements at grid station B. (a) Voltage THD and (b) current THD.

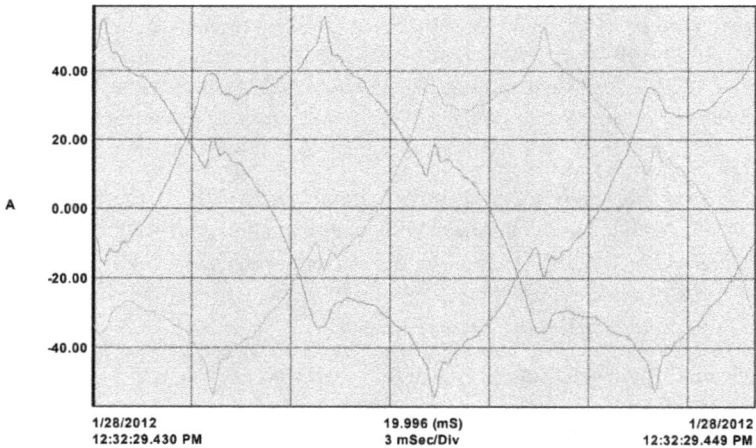

1/28/2012	19.996 (mS)	1/28/2012
12:32:29.430 PM	3 mSec/Div	12:32:29.449 PM

Figure 8. Current waveform of the 132 kV feeder at grid station B.

Using measurements, the maximum demand load current is 208 A. The short circuit level at grid station B is obtained from OETC capability statement. The TDD limit based on I_{SC}/I_L = 91 is 6%. A comparison between the 95th and 99th percentiles of current TDD and IEEE Std 519 limits is presented in **Table 8**.

Waveform I1
28.73 Arms, 18.88 %THD

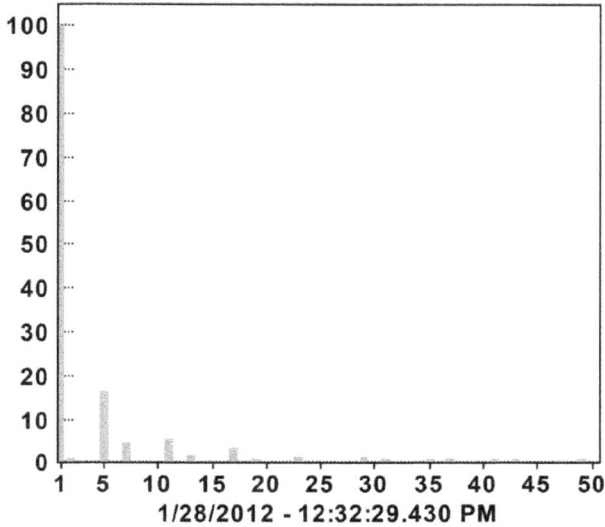

1/28/2012 - 12:32:29.430 PM

Figure 9. Grid station B current spectrum.

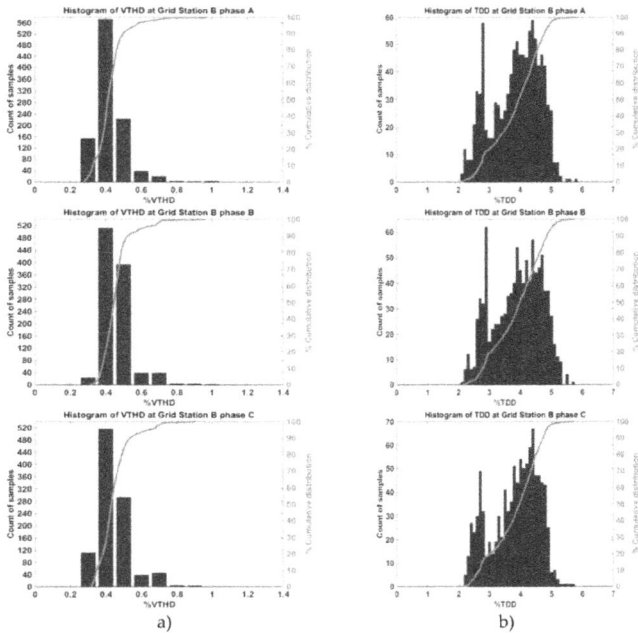

Figure 10. Histograms of THD measurements at grid station B. (a) Voltage THD and (b) current TDD.

Phase	THD P95	THD IEEE limits
A	0.58	2.5
B	0.621	2.5
C	0.641	2.5

Table 7. Voltage THD 95th percentile vs. IEEE Std 519-2014 at station B.

Phase	TDD P95	TDD P99	TDD IEEE limits
A	4.92	5.16	6
B	5.02	5.31	6
C	4.84	5.13	6

Table 8. Current TDD 95th and 99th percentiles vs. IEEE Std 519-2014 at station B.

4.2. Voltage harmonics temporal profile at grid station B

The voltage profile measured at grid station B is presented in **Figure 11** below. The daily diurnal profile is not clear due to high variability caused by load operations. These voltage fluctuations are attributed to fluctuation load current as seen in **Figure 12**.

Figure 13 presents the voltage THD as well as evident individual harmonics temporal profiles. It is worth nothing that no clear diurnal profile exists. Moreover, the dominant harmonic components are the 11th, the 13th, the 7th and the 5th. There is a small trace of the 3th and 9th harmonics. There is a very small trace of the 2nd harmonic as seen in **Figure 14**.

A similar observation can be seen in the 2nd location, where individual harmonic distortion is related to specific harmonic-producing loads. Station B is a grid station that supplies with proximity close to the largest industrial area in Oman. There are many non-linear loads fed from this station such as steel factories and aluminum smelters. The arc-furnaces loads, either induction furnaces or DC arc furnaces, along with static var compensators (SVCs) contribute significantly to the harmonic components of 11th, 13th, 7th, and 5th order.

4.3. Current harmonics temporal profile at grid station B

Figure 15 presents the current THD and individual harmonics profile at grid station B. It is worth noting that these profiles are highly fluctuating similar to measured current profile shown in **Figure 12**. It is worth mentioning that the dominant harmonic components are the 5th, the 11th, the 7th, and the 13th. In addition, there is a clear content of the 2nd, the 3rd, and other harmonic orders.

4.4. Reasons for temporal variations at grid station B

Comparing voltage temporal profile at grid station A with grid station B, it can be concluded that this variation depends on the type of loads available. While in grid station A, the variation

TIME PLOT - VOLTAGE CH1, 2, 3 Urms

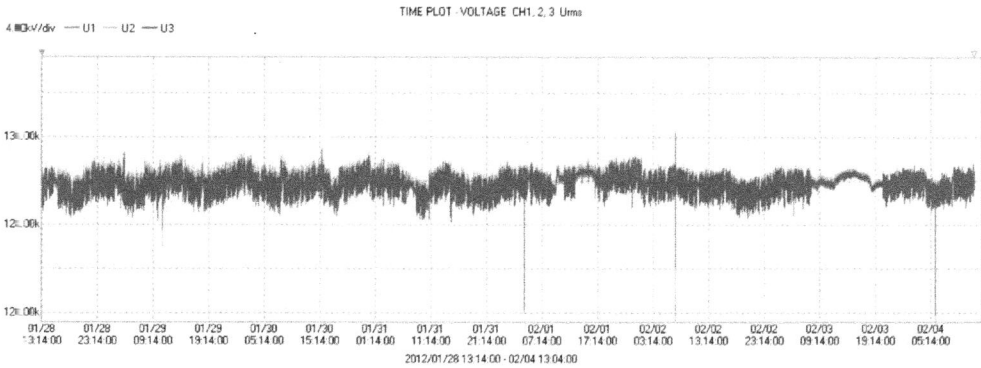

F gure 11. Voltage temporal profile at grid station B.

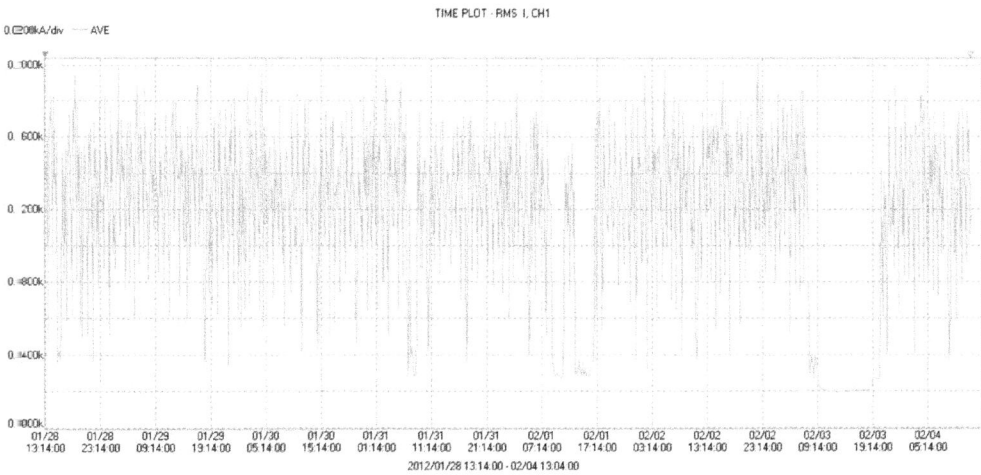

TIME PLOT - RMS I, CH1

F gure 12. Fluctuating load current at grid station B.

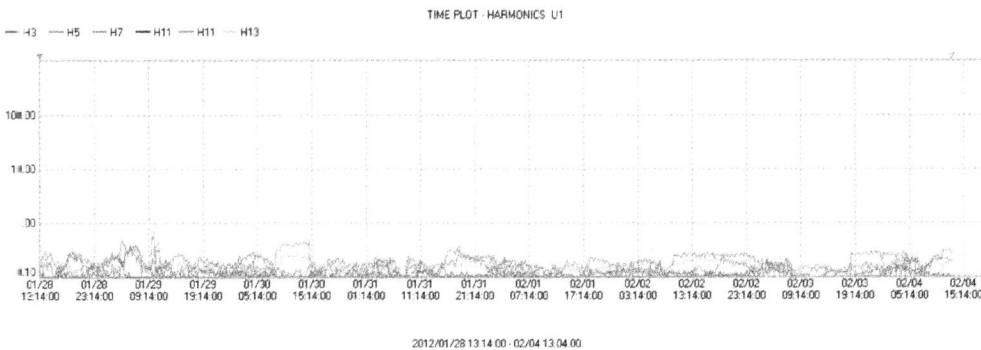

TIME PLOT - HARMONICS U1

2012/01/28 13 14 00 - 02/04 13.04 00

F gure 13. Voltage harmonics temporal profile at grid station B.

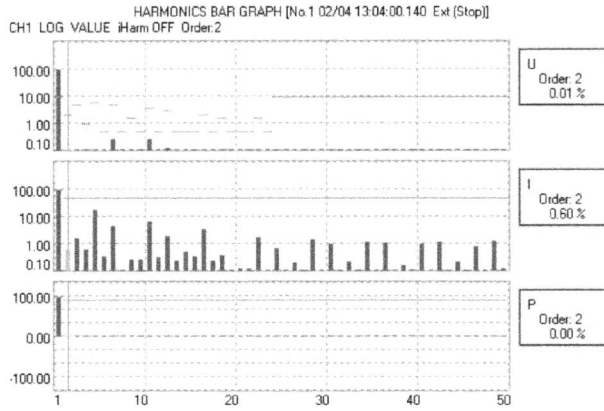

Figure 14. Harmonics bar graph at grid station B.

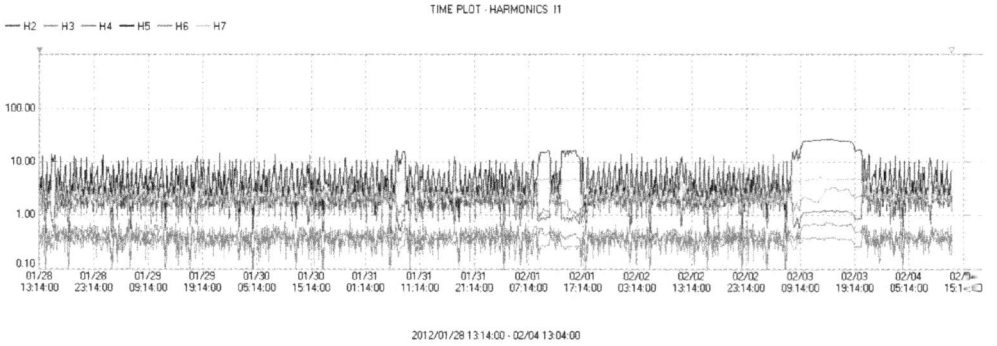

(a)

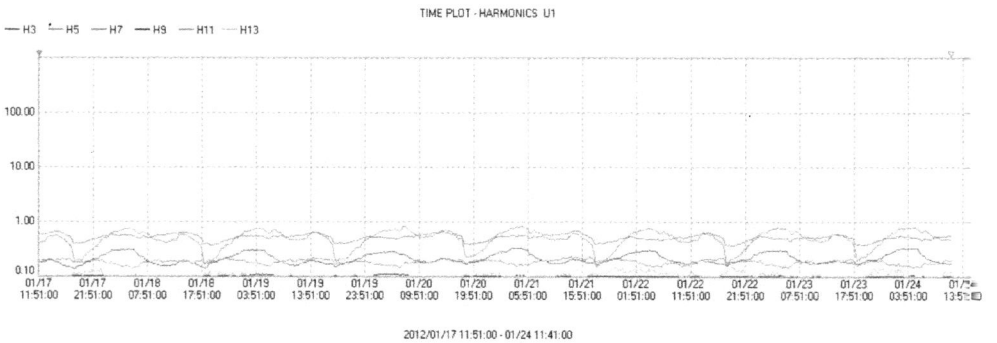

(b)

Figure 15. Current harmonics temporal profile at grid station B. (a) 7th and lower orders (b) 8th and higher orders.

reflects almost the daily load profile but in grid station B the variation is completely different owing to the existing of steel industries. Furthermore, although there is high current distortion in grid station B, the voltage THD is small due to the high short circuit capacity.

5. Conclusions

Harmonics were measured and analyzed for two grid stations in the main interconnected system of Oman. The first grid station is feeding both industrial and residential customers, while the second grid station is dedicated to large industrial customers including arc furnaces and rolling mills. A clear temporal variation of harmonics similar to that of aggregate load and local voltage profiles was observed at the grid station feeding both residential and industrial customers. However, this correlation between the system load and harmonics profiles diminishes at the grid station dedicated for heavy industrial loads.

Author details

Mohammed H. Albadi*, Rashid S. Al Abri, Amer S. Al Hinai and Abdullah H. Al-Badi

*Address all correspondence to: mbadi@squ.edu.om

Sultan Qaboos University, Muscat, Oman

References

[1] Khan S, Khan S, Ahmed G. Industrial Power Systems. USA: CRC Press; 2007

[2] Baggini A. Handbook of power quality, John Wiley & Sons; 2008. DOI: 10.1002/9780 470754245

[3] I. S. Association. 519-2014-IEEE Recommended Practices and Requirements for Harmonic Control in Electric Power Systems. New York: IEEE; 2014. DOI: 10.1109/IEEESTD. 2014.6826459

[4] IEC. "IEC 61000-3-6: Assessment of Emission Limits of Distorting Loads in MV and HV Power Systems," ed; 2008

[5] OETC. "The Grid Code, Version 2," ed; 2010

[6] MJEC, MZEC, and MEDC. "The Distribution Code, Version 1.000," ed; 2005

[7] OETC. "Five-Year Annual Transmission Capability Statement (2016-2020)"; 2016

[8] HIOKI. HIOKI E.E. CORPORATION. Hioki 3196 Power Quality Analyzer Meter. http://www.hioki.com/product/3196/3196v.html; 2014

[9]   Albadi M, Al Hinai A, Al-Badi A, Al Riyami M, Al Hinai S, Al Abri R. "Measurements and evaluation of harmonics in HV networks—Oman MIS case study". In: the 7th IEEE GCC Conference and Exhibition, Doha, Qatar, 17-20 November 2013. pp. 49-53. DOI:10.1109. IEEEGCC.2013.6705747

[10] Al-Badi A, Albadi M, Al-Hinai A. Designing of filter to reduce harmonic in industrial power networks. In: 3rd IEEE International Energy Conference (EnergyCon 2014) Dubrovnik, Croatia, 13-16 May 2014. DOI: 10.1109/ENERGYCON.2014.6850425

[11] Albadi M, Al Hinai A, Al-Badi A, Al Riyami M, Al Hinai S, Al Abri R. "Measurements and evaluation of flicker in high voltage networks". Renewable Energy and Power Quality Journal (RE&PQJ). 2014;**12**:1-6. DOI: 10.24084/repqj12.218

[12] Albadi M, Al Hinai A, Al-Badi A, Al Riyami M, Al Hinai S, Al Abri R. Unbalance in power systems: Case study. In: 2015 IEEE International Conference on Industrial Technology (ICIT), Seville, Spain, 2015. pp. 1407-1411. DOI: 10.1109/ICIT.2015.7125294

Chapter 4

Harmonic Distortion Caused by Single-Phase Grid-Connected PV Inverter

Yang Du and Dylan Dah-Chuan Lu

Additional information is available at the end of the chapter

http://dx.doi.org/10.5772/intechopen.73030

Abstract

Due to the fast growth of photovoltaic (PV) installations, concerns are rising about the harmonic distortion generated from PV inverters. A general model modified from the conventional control structure diagram is introduced to analyze the harmonic generation process. Causes of the current harmonics are summarized, and its relationship with output power levels is analyzed. In particular for two-stage inverter, unlike existing models that assume the direct current (DC)-link voltage is constant, the DC-link voltage ripple is identified as the source of a series of odd harmonics. The inverter is modeled as a time-varying system by considering the DC-link voltage ripple. A closed-form solution is derived to calculate the amplitude of the ripple-caused harmonics. The theoretical derivation and analysis are verified by both simulation and experimental evaluation.

Keywords: DC-link voltage ripple, harmonics, Matlab/Simulink, PV inverter, single phase

1. Introduction

Among numerous renewable energy sources, solar energy is considered as one of the most promising resources for large-scale electricity production [1]. In several countries including Australia, an increasing number of photovoltaic (PV) generation systems are connected to the distribution network as a result of strong government support. The PV market is growing rapidly (30–40%), and its price is constantly decreasing. Many countries are trying to increase the penetration of renewable energy.

The power electronics interface is essential for connecting renewable energy sources to the grid. This interface has two main functions such as extracting the maximum amount of power from the PV modules [2, 3] and conversion of direct current (DC) power to an appropriate form of alternative current (AC) power for the grid connection. Renewable energy sources such as solar energy cannot be manipulated in the same way as conventional power sources, so the operating conditions of PV inverters vary according to the solar insolation [4]. However, utility standards and manufacturers' data sheets are only concerned with the full-load condition.

PV systems incorporate power electronic interfaces, which generate a level of harmonics [5], potentially causing current and voltage distortions. The summations of various higher frequency sinusoidal components are the harmonics of current or voltage waveforms, which are an integer multiple of the fundamental frequency. These harmonics have a great influence on the operational efficiency and reliability of the power system, loads, and protective relaying [6]. Due to the rapid growth of PV installations, attention to harmonic distortion introduced by PV inverters to the grid is on the rise.

The degree of current total harmonic distortion (THD), as a ratio of the fundamental current and the real power output of the inverter, vary significantly [7]. At a low power output level, the current THD becomes higher, especially for generated power below 20% of the rated power, such as in the morning or evening. Many researchers have reported this phenomenon and tried to find out the causes. In the control system, the quantization and resolution effects of the measurement devices have been pointed out as one of the causes [8]. Another explanation is that the closed-loop current controls, which are intended to minimize the harmonic components, stop working at a low power output level [7]. Some researchers have suggested that the DC-link voltage regulation is highly related to the reference current resolution [9]. However, the comprehensive and systematic analysis of the generation process of the harmonics in the PV inverter output current is missing.

The conventional model of current control structure [10] is widely used to design the control loop and to analyze its stability. However, this model dose not including harmonic information, and the model cannot reflect the influence of the control schemes on the resulted harmonics. Section 2 introduces a general model modified from a conventional control structure diagram to analyze the harmonic generation process. The "harmonic impedance" concept [10] is used to quantitatively calculate the harmonic amplitude caused by each source. This is important because of the growing concern of harmonics generated by these devices and their effect upon other equipment.

A series of fund odd harmonics cannot be completely explained by the factors usually examined in such cases. These harmonics are caused by the DC-link voltage ripple, and a time-varying model is proposed to analyze this phenomenon in Section 4.

In order to analyze and design the PV inverter, the DC-link voltage is assumed as constant in the traditional model of a PV inverter. However, this is not always the case. The AC instantaneous output power exhibits a pulsation at the double-line frequency for single-phase grid-connected inverters. Under stable insolation conditions, the DC output voltage of the PV modules is controlled as constant at the maximum power point (MPP). Therefore, the power pulsation caused by single-phase power generation is converted into the static stored energy on the decoupling capacitor, and the double-line frequency of voltage ripple can be found at the DC link [11–13]. By using large electrolytic capacitors, the ripple can be reduced but not eliminated. However, the electrolytic parts have far more limited life than the applications [14] which need to be avoided.

A single-stage inverter is shown in **Figure 1(a)**; an efficient maximum power point tracking (MPPT) process is realized by a large power decoupling capacitor. Hence, modeling the inverter can use adaptable constant DC-link voltage assumption in this linear model. However, as two-stage inverter is shown in **Figure 1(b)**, the power decoupling capacitor is placed at the high-voltage DC link. In this topology, a larger voltage ripple is allowed to present across a DC link in order to minimize the decoupling capacitor [15], hence the constant DC-link voltage assumption is not valid.

The three-phase bridge converter for harmonic transfer is investigated in [16], the voltage second harmonic on a DC link producing a third harmonic on the AC side can be found. However, the DC-link voltage also causes output current frequency spectrum for the fifth, seventh, and a series of odd harmonics [17]. The explanation of this phenomenon cannot be found in the previous research. Many methods have been proposed to eliminate the current harmonics caused by the DC-link ripple without analyzing the harmonics generation process. A specifically designed pulse-width modulation (PWM) control algorithm [18] is proposed to compensate the DC-link voltage ripple. In [19], a control technique, which allows for 25% ripple voltage without distorting the output current waveform, has been proposed. The cutoff frequency of this design is 10 Hz, which could attenuate the voltage ripple in the control loop, but dynamic performance is decreased in this system. The main purpose of all these works is to eliminate the effects of the DC-link voltage ripple. However, an understanding of the relationship and the analytical model for qualitative information between the output current harmonics and DC-link voltage ripple is still missing.

In this chapter, for harmonic analysis studies, a new model of the single-phase full-bridge PV inverter is proposed by regarding its loading level and the ripple of the DC-side voltage. It is obtained by adding representation of the DC-link voltage ripple into the conventional linear model of a grid-connected PV inverter. Thus, it becomes a periodical time-varying model.

This chapter is organized as follows: a general model with harmonic information is introduced in Section 2. In Section 3, the double-line frequency voltage ripple on the DC link is identified as the cause of a series of odd harmonics. A time-varying model is proposed to analyze this phenomenon. Section 4 gives simulation and experimental results, which verify the validity of the proposed model and solution. Conclusions are given in Section 5.

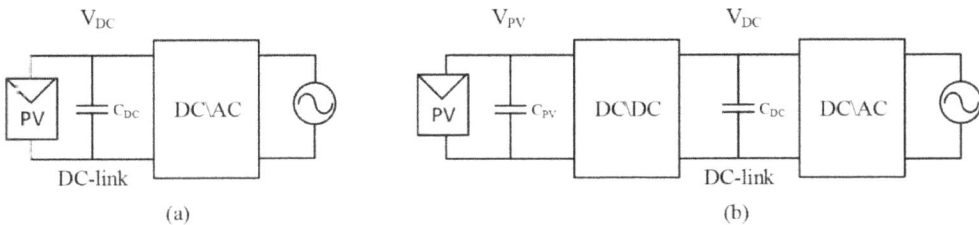

Figure 1. Block diagram of (a) single-stage inverter and (b) two-stage inverter.

2. Modeling of the PV inverter for harmonic analysis

In this section, PWM inverter framework with current feedback control for single-phase full-bridge PV inverter, which is generally used in the commercial products, conventional model of the current regulation scheme for that kind of inverter, and general inverter model proposed for the harmonic analysis are presented.

2.1. Single-phase full-bridge PV inverter with current control

An example of PWM inverter framework with current feedback control is shown in **Figure 2**. It is the most common structure which used by the commercial products. The inverter is formed by one output inductor, a DC-link capacitor C_{DC}, and four power switches. The DC-link voltage V_{DC} presents two different scenarios: one is with voltage ripple and another is without voltage ripple. The following sections analyze these two different cases separately. V_{inv} is the full-bridge inverter output voltage and V_g is the grid voltage, I_{out} is the inverter output current. A fixed grid voltage has been applied to the grid-connected inverter output terminals, and the inverter input voltage is controlled to provide MPP tracking. A current control scheme is used, since only the AC output current can be controlled. A filter has been used to connect between the inverter and the grid. In this chapter, a single inductor is used to simplify the analysis. A feedback control with the PI controller is used for the PWM inverter

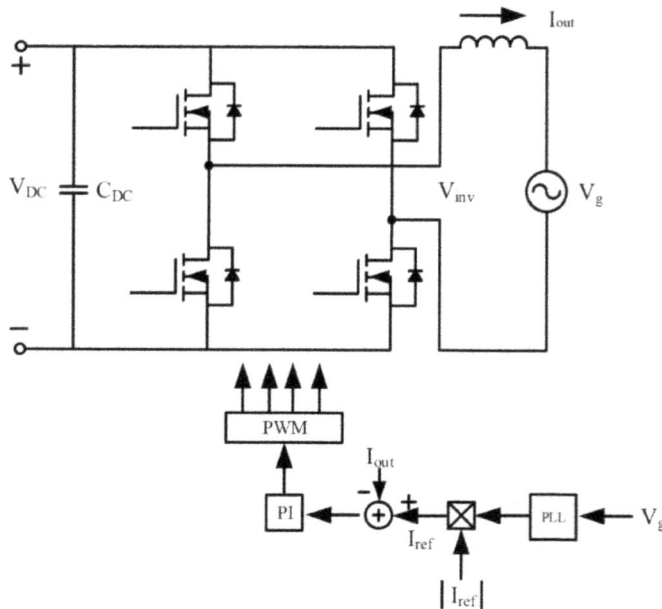

Figure 2. PWM inverter framework with current-controlled feedback loop.

to control the output current I_{out} to track a reference output current I_{ref}. A phase-locked loop (PLL) has been used to obtain the phase angle of I_{ref} from the grid voltage V_g. The amplitude of the reference current $|I_{ref}|$ can be determined by the voltage control loop according to the MPPT process. The design of the voltage control loop may vary according to different inverter topologies. The detailed derivation of $|I_{ref}|$ can be found in [10, 20].

2.2. Conventional model of current regulation scheme

Figure 3 shows the conventional control structure diagram of the current-controlled inverter. This model can be analyzed by using conventional linear analysis methods. It can help the designer to tune the controller [21] and investigate the control performance and stability [22]. The closed-loop transfer function is given by

$$I_{out} = \frac{G_{PI}G_{PWM}G_{inv}G_f}{1 + G_{PI}G_{PWM}G_{inv}G_f}I_{ref} - \frac{G_f}{1 + G_{PI}G_{PWM}G_{inv}G_f}V_g \tag{1}$$

where G_{PI}, G_{PWM}, G_{inv}, and G_f are the transfer functions for the PI controller, PWM, inverter, and filter, respectively. In this model, only the fundamental waveforms are considered, and harmonic information is required for the harmonic distortion analysis.

2.3. The general inverter model for the harmonic analysis

Figure 4 shows the generalized model which is derived from the conventional current control structure diagram for a PWM inverter with harmonic information. The location and types of harmonic sources, which need to be added, are shown in this figure. The output current $S5$ is generated based on a reference current by the full-bridge inverter with current control, as shown in the first trace. The model of current control scheme, which includes the harmonics information, is shown in the second trace. Compared with **Figure 3**, the switch harmonic source $V_{switch\ harmonics}$ in the PWM section is added to generate a pulse waveform on top of the sinusoidal signal. This harmonic source contains the characteristic of the PWM, including the type of PWM method and the switching frequency. The voltage difference between the grid voltage and the inverter output voltage will cause the changes in the output current. Therefore, in **Figure 4**, the grid voltage harmonic source is added at the inverter

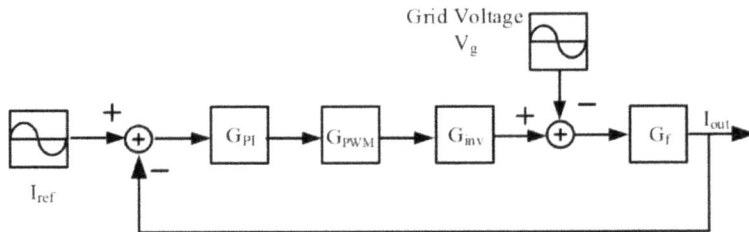

Figure 3. Conventional control structure diagram of the current-controlled inverter.

section. The lowest trace in **Figure 4** sketches the waveform of each stage, and the details are described as follows:

1. $S1$ is the error between the current reference and the output current of the inverter, $S1 = I_{ref} - I_{out} = I_{ref} - S5$.

2. $S2$ is the amplitude modulation (AM) ratio, $S2 = S1G_{PI}$, where the PI controller's transfer function is $G_{PI} = k_p + k_i/s$. k_p and k_i are the proportional and the integral gain.

3. $S3$ is the gate drive signal. $S3 = S2G_{PWM} + V_{switch\ harmonics}$, where $G_{PWM} = 1/C_{pk}$ and C_{pk} is the carrier signal's peak value.

4. $S4$ is the output voltage of the inverter V_{inv}, $S4 = S3G_{inv}$, where $G_{inv} = V_{DC}$. The V_{DC} can be either a time-varying or a constant signal; these two cases need to be treated separately.

5. $S5$ is the output current of the inverter I_{out}. $S5 = (S4 - V_g)G_f$. The grid voltage V_g may contain the voltage harmonics $V_{g\ harmonics}$. $S4 - V_g$ is the voltage difference between the output filter. The transfer function of the filter is $G_f = 1/(Ls)$, where L is the filter's inductance.

The main causes of harmonic in PV inverter can be summarized into several categories: grid background voltage distortion, switch harmonics (high frequency), DC-link voltage variation due to MPPT, and some other causes (PLL blocks, etc.). Harmonic distortion for both cases, with or without voltage ripple on the DC link, can be analyzed by using this generalized model.

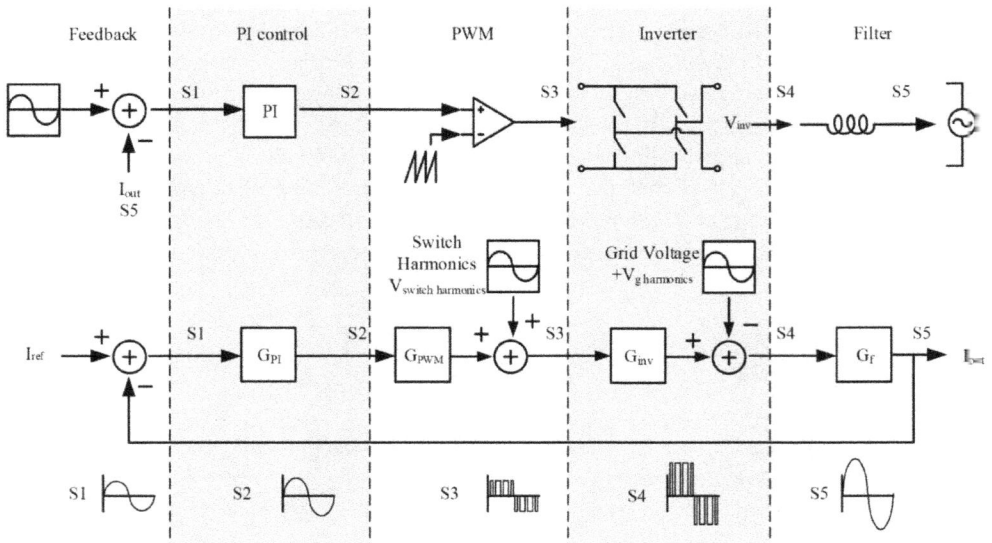

Figure 4. Model of current-controlled PWM inverter with harmonic information.

3. Current harmonic caused by DC-link voltage ripple

In this section, the current harmonics caused by DC-link voltage ripple has been analyzed. The model for considering the double-line frequency voltage ripple has been built. The closed-form solution for the current harmonics has been provided.

Figure 5 shows the model of the inverter based on **Figure 4**, and the DC-link voltage ripple has been taken into account. The inverter transfer function G_{inv} shown in **Figure 4** is replaced by the section under the triangle shading, which is a sinusoidal signal V_{rip} at double-line frequency on top of the DC component V_{DC}. Since the voltage ripple is time-varying, the transfer function for this section cannot be derived. In [23], the authors point out that closed-form solutions cannot be derived when the harmonic ripple components are not neglected. However, numeric solutions can be evaluated for any particular operating condition by using this model.

The harmonic characteristics of the output current shown in **Figure 5** can be identified by qualitatively analyzing the simplified loop model. The section under the triangle shading is also known as the amplitude modulation; the feedback loop with unit delay is shown in **Figure 6**, where Z^{-1} denotes the delay of a unit sample period. Compared with **Figure 6**, in this simplified model, several linear blocks are left out. Due to the system linearity, the signal frequency characteristics will remain the same. A similar analysis method, which has been used in sound processing research [24], is adopted in this chapter to analyze this time-varying system. Two discrete-time sinusoidal example signals $I_{ref}[n] = \cos(\omega_o n)$ and $V_{rip}[n] = \cos(2\omega_o n)$ are used. The output signal $y[n]$ can be illustrated as the result of subtraction between the reference signal $\cos(\omega_o n)$ and the delayed output signal $y[n-1]$ then timed with the AM section, which is $\cos(2\omega_o n) + V_{DC}$

$$
\begin{aligned}
y[n] &= [\cos(\omega_o n) - y[n-1]][\cos(2\omega_o n) + V_{DC}] \\
&= \cos(\omega_o n)[\cos(2\omega_o n) + V_{DC}] - y[n-1][\cos(2\omega_o n) + V_{DC}]
\end{aligned}
\tag{2}
$$

For $n \leq 0$, ω_o is the angular velocity of a signal in the fundamental frequency, V_{DC} is constant, at the initial condition, $y[n] = 0$,. The delay exists at any point in time n, and we need to store $y[n-1]$ so that it can be used in the computation of $y[n]$. The $y[n-1]$ is

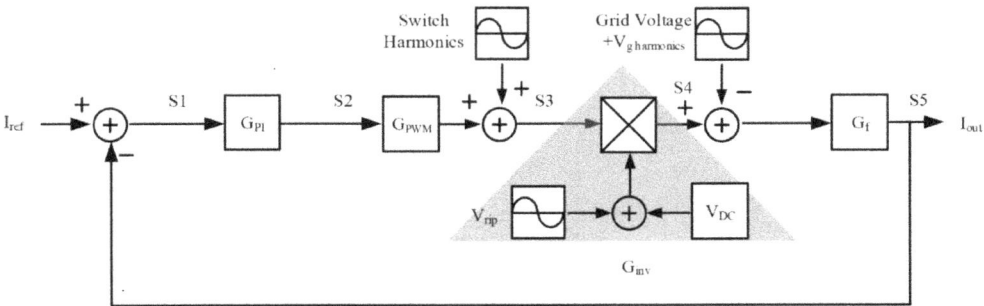

Figure 5. Model of inverter with the DC-link voltage ripple.

$$y[n-1] = [\cos(\omega_o[n-1]) - y[n-1-1]][\cos(2\omega_o[n-1]) + V_{DC}]$$
$$= \cos(\omega_o[n-1])[\cos(2\omega_o[n-1]) + V_{DC}] - y[n-2][\cos(2\omega_o[n-1]) + V_{DC}]$$

(3

Substitute Eq. (3) into Eq. (2)

$$y[n] = \cos(\omega_o n)[\cos(2\omega_o n) + V_{DC}]$$
$$- \cos(\omega_o[n-1])[\cos(2\omega_o[n-1]) + V_{DC}][\cos(2\omega_o n) + V_{DC}]$$
$$+ y[n-2][\cos(2\omega_o[n-1]) + V_{DC}][\cos(2\omega_o n) + V_{DC}]$$

(4

This feedback expression can be expanded into an infinite summation of products given by

$$y[n] = \cos(\omega_o n)[\cos(2\omega_o n) + V_{DC}]$$
$$- \cos(\omega_o[n-1])[\cos(2\omega_o[n-1]) + V_{DC}][\cos(2\omega_o n) + V_{DC}]$$
$$+ \cos(\omega_o[n-2])[\cos(2\omega_o[n-2]) + V_{DC}][\cos(2\omega_o[n-1]) + V_{DC}][\cos(2\omega_o n) + V_{DC}]$$
$$- \cos(\omega_o[n-3])[\cos(2\omega_o[n-3]) + V_{DC}][\cos(2\omega_o[n-2]) + V_{DC}][\cos(2\omega_o[n-1])$$
$$+ V_{DC}][\cos(2\omega_o n) + V_{DC}]$$

(5

A series of odd harmonics is caused by this amplitude modulation in a feedback loop. In Eq. (5), the first product term is illustrated as an example. According to Euler's formula, this term can be expressed as the sum of sinusoids with angular velocity ω_o and $3\omega_o$, which is the fundamental and third harmonics.

$$y[n] = \cos(\omega_o n)[\cos(2\omega_o n) + V_{DC}] = \cos(\omega_o n)V_{DC} + \frac{1}{2}\cos(3\omega_o n) + \frac{1}{2}\cos(\omega_o n)$$

(6

The closed-form solution is derived based on an idea which is similar to the harmonic balance method for radio frequency (RF) circuit [25]. The harmonic balance method is a frequency domain method for calculating steady states of a nonlinear circuit.

All the signals in the control loop can be expressed in polar forms by taking a Fourier transform. The high-order harmonics will be attenuated by the feedback loop, and only the low-order harmonics will be considered. The low-pass filter (LPF) can be assumed as an ideal filter, which can eliminate all the harmonics above a certain order. A finite number of equations can be

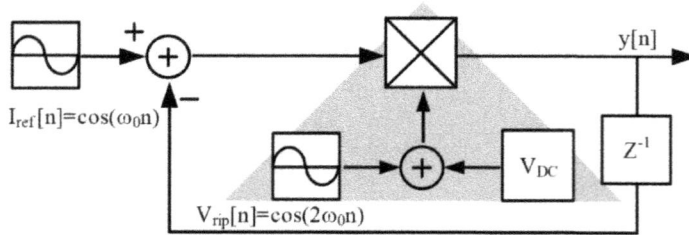

Figure 6. Amplitude modulation in unit delay feedback.

obtained by using this assumption, and a partial linearization for the control loop can be achieved. The amplitude of the feedback signals can be set as variables. After equating like terms in the equation of the output current and in the preset amplitude of feedback signal, a number of equations can be obtained. By solving these equations, the closed-form solution for the amplitude of a certain order of harmonic can be derived. Detailed analysis of the simplified model by using this method in a time domain is given as follows:

The output signal $y(t)$ of the simplified model is the difference between the current reference $I_{ref}(t)$ and the feedback signal $I_{fb}(t)$ multiplied by the DC-link voltage

$$y(t) = \left(I_{ref}(t) - I_{fb}(t)\right)\left(V_{rip}(t) + V_{DC}\right) \tag{7}$$

The signals in **Figure 6** can be expressed in polar form

$$I_{ref}(t) = A\cos(\omega_o t) = ae^{j\omega_o t} + ae^{-j\omega_o t}, a = \frac{1}{2}A \tag{8}$$

$$V_{rip}(t) = B\cos(2\omega_o t) = be^{j2\omega_o t} + be^{-j2\omega_o t}, b = \frac{1}{2}B \tag{9}$$

where A and B are the amplitudes of $I_{ref}(t)$ and $V_{rip}(t)$. In order to simplify the analysis to a level that is suitable for manual calculation, the feedback signal $I_{fb}(t)$ is assumed to include only fundamental and third harmonics

$$I_{fb}(t) = C_1 \cos(\omega_o t) + C_3 \cos(3\omega_o t) = c_1 e^{j\omega_o t} + c_1 e^{-j\omega_o t} + c_3 e^{j3\omega_o t} + c_3 e^{-j3\omega_o t} \tag{10}$$

where C_1 and C_3 are the assumed variables for the amplitude of the fundamental component and third harmonics, and $c_1 = 0.5C_1, c_3 = 0.5C_3$. This can be easily extended to any number of harmonics with the help of computer-aided calculations.

Eq. (11) can be obtained by substituting Eqs. (8)–(10) into Eq. (7). Since the $I_{fb}(t)$ is the low-order part of y(t), the harmonic amplitude equation can be found by equating like terms in Eqs. (10) and (11)

$$y(t) = [(V_{DC} + b)(a - c_1) - bc_3]e^{j\omega_o t} + [(V_{DC} + b)(a - c_1) - bc_3]e^{-j\omega_o t}$$
$$+ [b(a - c_1) - V_{DC}c_3]e^{j3\omega_o t} + [b(a - c_1) - V_{DC}c_3]e^{-j3\omega_o t} - bc_3 e^{j5\omega_o t} - bc_3 e^{-j5\omega_o t} \tag{11}$$

$$\begin{cases} c_1 = (V_{DC} + b)(a - c_1) - bc_3 \\ c_3 = b(a - c_1) - V_{DC}c_3 \end{cases} \tag{12}$$

All the parameters are fixed values (a, b and V_{DC}) for a specific inverter; therefore, the harmonic amplitude can be obtained by substituting these values. By using the same method, the closed-form solution for the averaged model in **Figure 5** with PI controller can be derived. Two integral sections will be involved into the calculation due to the integrator in the controller and the filter. This calculation for the practical models becomes significantly complicated, and it is impossible to calculate manually. Matlab can be utilized as an effective tool to conduct these calculations.

4. Simulation and experimental results

In this section, the simulation and experiment results are reported. By using Matlab/Simulink a single-phase PV inverter is simulated. The switching model simulation provides the most detailed results including the switch information and all the potential harmonic distortions. By using the Simulink SimPowerSystems toolbox, the developed model includes both electronic components and control blocks. The schematic diagram is shown in **Figure 7**. It is a time-consuming process. A 0.2-s simulation period takes about 10 min to run on a computer with average performance (Intel Core 2 Duo CPUs and 4 GB of 800 MHz DDR2 RAM).

The two-stage grid-connected PV system prototype is constructed in the laboratory to verify the abovementioned analysis. It includes a boost converter connected with the full-bridge inverter as the second stage. The operating voltage range of the system has been scaled down due to the limitation in the experimental setup. An AC source with 50 V_{rms} is used as the grid voltage. The full-bridge inverter DC-link voltage is 100 V. The capacitor size has been changed with different capacitance to create different voltage ripple across the DC link. The experimental setup with the prototype circuit is shown in **Figure 8**, which is the same as in **Figure 9**, and the main circuit parameters are shown in **Table 1**. A dSPACE controller set has been used to control these two stages.

An averaged model has also been built in Simulink. The parameters listed in **Table 1** are substituted into the derived closed-form solution in order to calculate the harmonics. Different levels of DC-link voltage ripple in simulation and experiment have been created by using different sizes of capacitors. Only the third-order harmonics from the fast Fourier transform (FFT) analysis is considered in calculating the THD, in order to simplify the analysis process. In

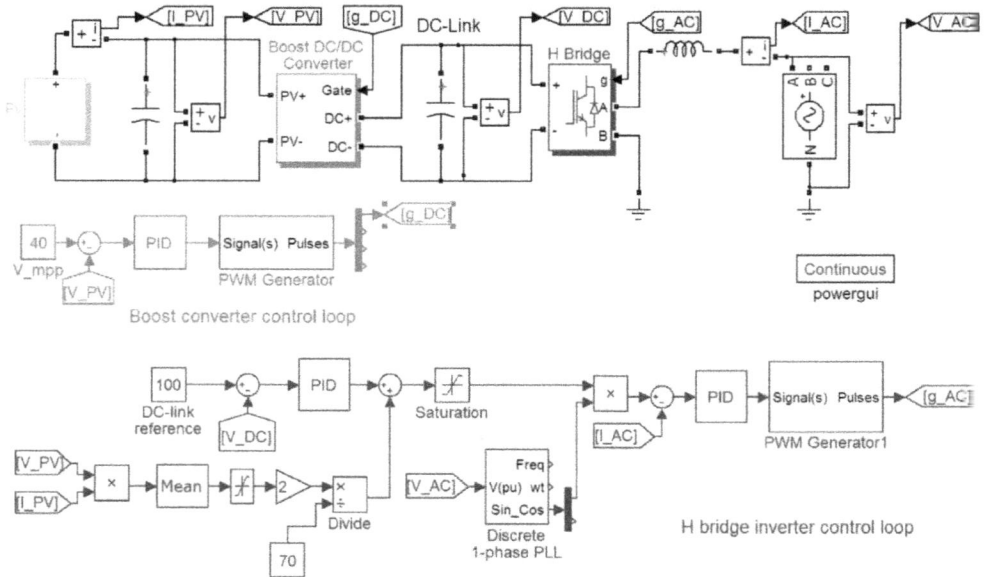

Figure 7. Simulation schematic model.

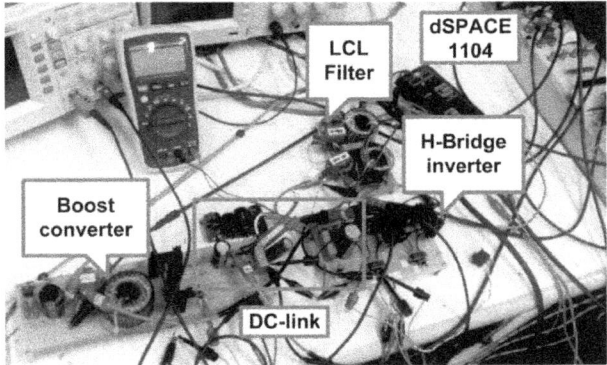

Figure 8. Experimental setup.

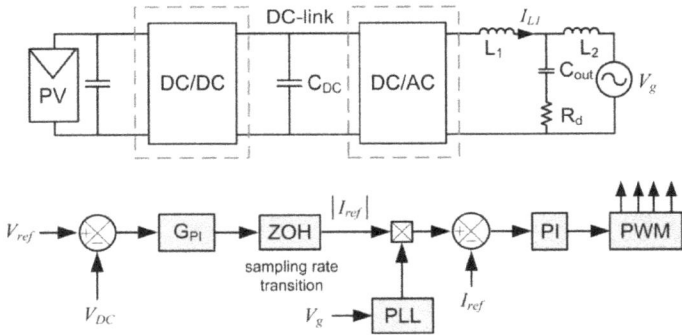

Figure 9. Two-stage PV inverter with feedback control.

Parameter	Label	Value	Unit
Switching frequency	f_{sw}	20	kHz
Rated output frequency	f	50	Hz
Rated output voltage	V_g	70	V
DC-link capacitance	C_{DC}	770	μF
DC-link voltage	V_{DC}	100	V
Inverter-side inductor	L_1	2.56	mH
Grid-side inductor	L_2	1.10	mH
Output capacitor	C_{out}	2.2	μF
Damping resistor	R_d	1	Ω

Table 1. Specification of the PV inverter.

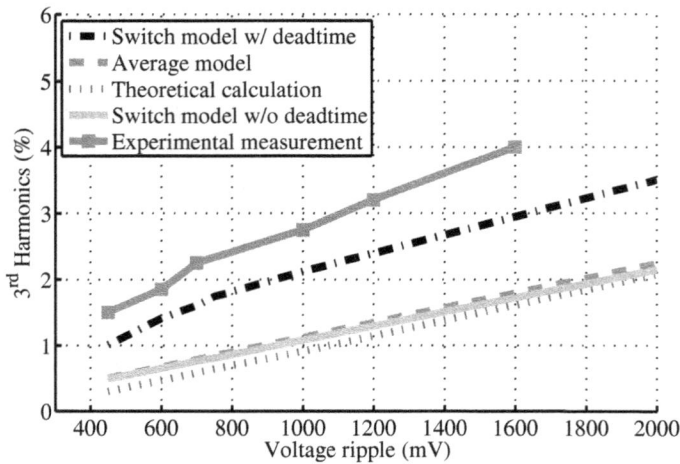

Figure 10. Simulation, calculation, and experimental results.

Figure 10, the switch model, averaged model, and calculation results agree with one another. It proves that the closed-form solution can fully represent the switch model for harmonic analysis. Higher-order harmonics also can be analyzed by using the same method.

The same parameters as the simulation have been used in the experiment, and the results are also plotted in **Figure 10**. The experimental results show the same trend as the analysis suggests that the harmonic distortion increases as the DC-link voltage increases.

5. Conclusion

In this chapter, a general model, which is modified from a conventional control structure diagram, has been introduced to analyze the harmonic generation process caused by single-phase PV inverter. The causes of the harmonics have been identified. A series of odd harmonics in the output current on the DC-link capacitor are generated by the double-line frequency voltage ripple. In this chapter, a nonlinear, time-varying model and its closed-form solution were provided. The relationship between the amplitude of the harmonics and the DC-link voltage ripples has been presented. The proposed solutions are proved by both experimental and simulation results. It is a tool for evaluating the power-quality issues in grid-connected inverter systems. The designers can also use it to consider the tradeoff between the size of the DC-link capacitor and the output harmonics in the output current.

Acknowledgements

This research was supported by the University Research Development Fund (RDF-15-01-40) from the Xi'an Jiaotong-Liverpool University and Jiangsu University S&T programme (17KJB470012).

Author details

Yang Du[1]* and Dylan Dah-Chuan Lu[2]

*Address all correspondence to: yang.du@xjtlu.edu.cn

1 Xi'an Jiaotong-Liverpool University, Suzhou, China

2 University of Technology Sydney, Australia

References

[1] Du Y, Lu DDC, Cornforth D, James G. A study on the harmonic issues at CSIRO microgrid. IEEE 9th International Conference on Power Electronics and Drive Systems (PEDS). 2011:203-207

[2] Du Y, Lu DD-C. Battery-integrated boost converter utilizing distributed MPPT configuration for photovoltaic systems. Solar Energy. 2011;**85**(9):1992-2002

[3] Bennett T, Zilouchian A, Messenger R. Photovoltaic model and converter topology considerations for MPPT purposes. Solar Energy. 2012;**86**(7):2029-2040

[4] Lu DDC, Nguyen QN. A photovoltaic panel emulator using a buck-boost DC/DC converter and a low cost micro-controller. Solar Energy. 2012;**86**(5):1477-1484

[5] Papaioannou IT, Alexiadis MC, Demoulias CS, Labridis DP, Dokopoulos PS. Modeling and measurement of small photovoltaic systems and penetration scenarios. Power Tech conference, Bucharest. 2009:1-7

[6] Jain SK, Singh SN. Harmonics estimation in emerging power system: Key issues and challenges. Electric Power Systems Research. 2011;**81**(9):1754-1766

[7] Chicco G, Schlabbach J, Spertino F. Experimental assessment of the waveform distortion in grid-connected photovoltaic installations. Solar Energy. 2009;**83**(7):1026-1039

[8] Infield DG, Onions P, Simmons AD, Smith GA. Power quality from multiple grid-connected single-phase inverters. IEEE Transactions on Power Delivery. 2004;**19**(4):1983-1989

[9] Wu TF, Sun KH, Kuo CL, Yu GR. Current distortion improvement and dc-link voltage regulation for bi-directional inverter in dc-microgrid applications. Applied Power Electronics Conference and Exposition (APEC). 2011:1582-1587

[10] Twining E, Holmes DG. Grid current regulation of a three-phase voltage source inverter with an LCL input filter. IEEE Transactions on Power Electronics. 2003;**18**(3):888-895

[11] Shimizu T, Wada K, Nakamura N. Flyback-type single-phase utility interactive inverter with power pulsation decoupling on the DC input for an AC photovoltaic module system. IEEE Transactions on Power Electronics. 2006;**21**(5):1264-1272

[12] Du Y, Lu DD-C, James G, Cornforth DJ. Modeling and analysis of current harmonic distortion from grid connected PV inverters under different operating conditions. Solar Energy. 2013;**94**(0):182-194

[13] Du Y, Lu D, Chu G, Xiao W. Closed-Form Solution of Time-Varying Model and its Applications for Output Current Harmonics in two-Stage PV Inverter. IEEE Transaction on Sustainable Energy; 2015

[14] Lahyani A, Venet P, Grellet G, Viverge PJ. Failure prediction of electrolytic capacitors during operation of a switch mode power supply. IEEE Transactions on Power Electronics. 1998;**13**(6):1199-1207

[15] Hu H, Harb S, Kutkut N, Batarseh I, Shen ZJ. Power decoupling techniques for micro inverters in PV systems-a review. IEEE Energy Conversion Congress and Exposition (ECCE). 2010:3235-3240

[16] Jiang Y, Ekstrom A. General analysis of harmonic transfer through converters. IEEE Transactions on Power Electronics. 1997;**12**(2):287-293

[17] Wang F, Duarte JL, Hendrix MAM, Ribeiro PF. Modeling and analysis of grid harmonic distortion impact of aggregated DG inverters. IEEE Transactions on Power Electronics. 2011;**26**(3):786-797

[18] Enjeti PN, Shireen W. A new technique to reject DC-link voltage ripple for inverters operating on programmed PWM waveforms. IEEE Transactions on Power Electronics. 1992;**7**(1):171-180

[19] Brekken T, Bhiwapurkar N, Rathi M, Mohan N, Henze C, Moumneh LR. Utility-connected power converter for maximizing power transfer from a photovoltaic source while drawing ripple-free current. In: IEEE 33rd Annual Power Electronics Specialist Conference (PESC). 2002. pp. 1518–1522 vol.3

[20] Zhao W, Lu DDC, Agelidis VG. Current control of grid-connected boost inverter with zero steady-state error. IEEE Transactions on Power Electronics. 2011;**26**(10):2825-2834

[21] Armstrong M, Atkinson DJ, Johnson CM, Abeyasekera TD. Low order harmonic cancellation in a grid connected multiple inverter system via current control parameter randomization. IEEE Transactions on Power Electronics. 2005;**20**(4):885-892

[22] Maknouninejad A, Kutkut N, Batarseh I, Zhihua Q, Shoubaki E. Detailed analysis of inverter linear control loops design. Applied Power Electronics Conference and Exposition (APEC). 2011:1188-1193

[23] McGrath BP, Holmes DG. A general analytical method for calculating inverter dc-link current harmonics. IEEE Transactions on Industry Applications. 2009;**45**(5):1851-1859

[24] Kleimola J, Lazzarini V, Valimaki V, Timoney J. Feedback amplitude modulation synthesis. EURASIP Journal on Advances in Signal Processing. 2011;**2011**(1):434378

[25] Maas SA. Nonlinear Microwave and RF Circuits: Artech House Publishers; 2003

Chapter 5

New Trends in Active Power Filter for Modern Power Grids

Luís Fernando Corrêa Monteiro

Additional information is available at the end of the chapter

Http://dx.doi.org/10.5772/intechopen.72195

Abstract

From harmonic compensation to interface with renewable energy sources, active filters are capable to improve power quality, increase the reliability of the power grid, and contribute to make feasible the implementation of decentralized microgrids. In this scenario, this chapter provides a discussion involving new trends on distribution power grids, with active power filters playing an important key role. Considering the afore-mentioned explanation, part of the chapter covers active filter applications for power grids. In sequence, we discuss time domain control algorithms to identify power quality disturbances or other problems that may compromise the power grid reliability, with simulation results to evaluate the performance of the active filters for compensating power quality problems under transient- and steady-state conditions. Next, we discuss the integration of active filters with renewable energy sources (RENs) including a brief explanation of maximum power point tracking (MPPT) algorithms and other controllers considering a decentralized microgrid scenario with several active filters connected at the same grid circuit.

Keywords: active filters, current and voltage compensation, real-time algorithms, renewable energy sources, microgrids

1. Introduction

In this section, we present a discussion involving basic aspects of the active filters for generation and distribution grids. It is important to comment that there are also power electronics compensators for transmission grids presenting different features as, for instance, damping subsynchronous resonance [1], power flow control [2, 3] and improve the stability of a power system [4]. These compensators are known as Flexible AC Transmission System (FACTS), and their study is beyond the scope of this chapter.

Backing to the active power filters, they can be understood as a controlled current sources or controlled voltage sources capable for compensating different power quality problems as, for

IntechOpen

instance, harmonic and unbalanced components, power factor, voltage sags or swells, damping low-frequency harmonic oscillations, and so on [5, 6]. Moreover, they are used as an interface for renewable energy sources in a new concept of distributed generation or even making the implementation of decentralized microgrids reliable [7–9].

A simplified scheme of the shunt active filter compensating all the harmonic currents drawn by the load is illustrated in **Figure 1**. An active filter is comprehended by power and control stages. The power stage comprises a voltage source converter (VSI), with a storage energy element (capacitor) at its DC link, inductor filter (L_{fp}), and small passive filters (Z_{fp}) to provide a low impedance path to the high-frequency components of the produced current by the VSI (i_{Lfp}). The control stage presents measurement and instrumentation circuits, microcontrollers, and VSI drivers. As indicated in **Figure 1**, the reference current produced by the VSI (i^*) is determined based on the applied control algorithms, which presents the load current (i_L), grid voltage (v_S) and the DC-link voltage (v_{DC}) as inputs. There is also a pwm controller for keeping i_{Lfp} in conformity with the reference current (i^*). A common point (**cp**) was considered to indicate that in a three-phase circuit, the passive filters are connected at this point of the circuit.

It is important to comment that an inductor (L_S) is usually applied to represent the grid impedance, which reflects the inductance characteristics of line cables and power transformers. Nevertheless, current researches point out to replace its representation by equivalent imped-ances that are dynamically modified due to a considerable amount of nonlinear loads, which are dynamically connected and removed from the power grid. This issue becomes more important nowadays due to the proliferation of renewable energy sources with power con-verter interface [10–13].

Figure 2 illustrates a simplified scheme of the series active filter compensating harmonics and voltage sag, with the reference voltage (v^*) being determined through the applied control algorithms, which presents the grid current (i_S), grid voltage (v_S), and the DC-link voltage (v_{DC}) as inputs. Moreover, there is a pwm controller for producing the VSI filtered voltage (v_{Zsf}).

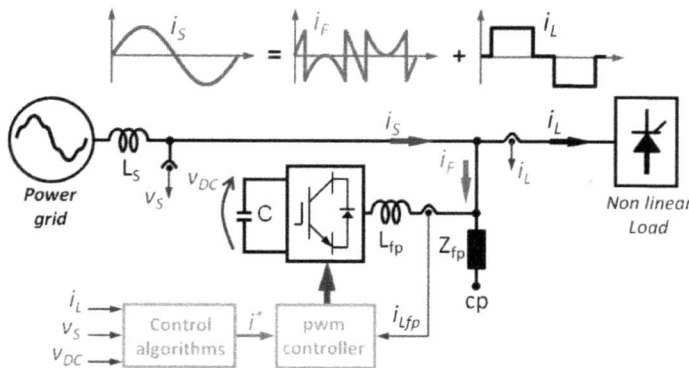

Figure 1. Simplified scheme of the shunt active filter compensating all the harmonic currents drawn by the nonlinear load.

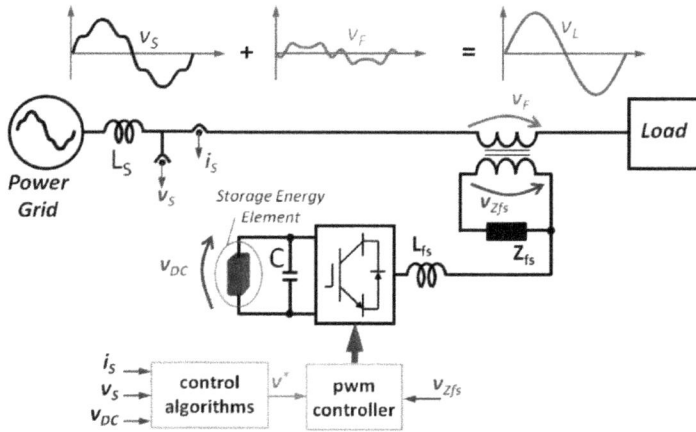

Figure 2. Simplified scheme of the series active filter compensating harmonics and voltage sag.

An additional storage energy element (SEE) is necessary if sag compensation is required. As depicted **Figure 2** with a SEE connected in parallel with the DC-link capacitor, it can be represented as, for instance, ultracapacitors or batteries [14]. There are also other SEEs as superconducting magnetic energy storage (SMES) [15] and flywheel [16]. However, once they are not voltage-source type, it is necessary to use power converters to interface them with the DC-link voltage.

Other issue involving the power stage of the series active filter corresponds to its series connection, which may or may not be done through power transformers. A constraint for implementing active filters without series transformer injection is to avoid short circuits between the phase circuits, which can be done replacing the three-phase VSI by three single-phases VSIs with three independent DC-link voltages as introduced in [17, 18]. Other alternative is the use of high-frequency transformers at the DC-link of the single-phase VSIs, which are usually applied in isolated DC-DC converters [19].

Other possible active filter topology consists on the combination of the shunt and series active filters, resulting on the unified power quality conditioner (UPQC). As described in [20], by having these two conditioners connected to the electrical system, simultaneous compensation of the current demanded from the utility and the voltage delivered to the load can be accomplished. As illustrated in **Figure 3**, the series and shunt active filters compensate at the same time all the harmonic components of the load currents and grid voltages. Its power stage combines all the passive components of the series and shunt active filters as previously exploited. In the same way, its control stage presents all the circuitry, microcontrollers, and control algorithms of both active filters. Particularly, in this configuration, the shunt active filter is responsible to draw a controlled current to keep the DC-link voltage (v_{DC}) regulated. A summary of the UPQC compensation capabilities is shown in **Table 1**, with the functionalities of the series and shunt active filters well established. Nevertheless, there are proposals in the literature with both active filters compensating the same power quality problem in a complementary way. For instance [21, 22]

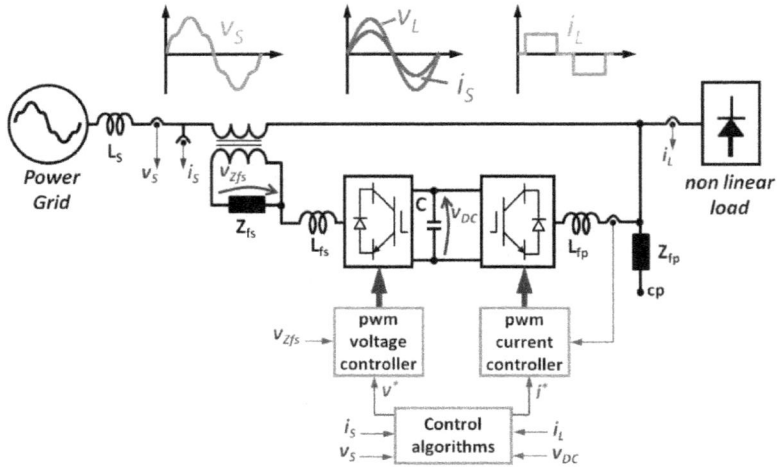

Figure 3. Simplified scheme of the unified power quality conditioner (UPQC).

Active filter	Functionality
Series active filter	Harmonic and unbalanced voltages compensation
	Voltage sags/swells compensation
	Improvement of the power grid stability
Shunt active filter	Harmonic and unbalanced currents compensation
	Power factor correction

Table 1. UPQC functionalities.

present a sag compensation proposal through the combined operation of the series and shunt active filters for the maximum utilization of both active filters.

In next section, we exploit their control algorithms.

2. Overview of active power filters for current and voltage compensation

Due to the power grid dynamics, an instantaneous or, at least, a quasi-instantaneous response of the active filters is desirable, which leads the use of time domain control algorithms together with synchronizing circuits. Hence, in this section, we exploit control algorithms to the series and shunt-active filters with simulation results.

2.1. Control algorithms for shunt active filter

Basically, control algorithms for shunt active filters can be divided into a set of algorithms for determining the reference current and other algorithms for controlling the produced current by the VSI, which depends on the applied switching technique.

Algorithms for determining the reference current are related to which features we expect that the active filter be able to compensate. It is important to comment that control algorithms for shunt active filters have been proposed in the literature for more than 30 years. Among all these proposals, those derived from the instantaneous power theory [23–25], dq reference frame [26–28], conservative power theory [7], and the active and non-active currents [29–31] are widely applied.

The instantaneous power theory, or p-q theory, was emerged at the beginning of the 1980s, with the main purpose to provide new power definitions in time domain for three-phase three-wire circuits and, in sequence for three-phase four-wire circuits. Based on the $\alpha\beta0$ system coordinates, the p-q theory has the advantage of instantaneously separating the homopolar (zero-sequence) from the nonhomopolar (positive- and negative-sequence) components [31]. This issue allowed new proposals on control algorithms to three-phase four-wire active filters. An enhanced version of the p-q theory, known as the *p-q-r* theory, was conceived based on a different coordinate translation, where voltages and currents are translated from $\alpha\beta0$ to *p-q-r* system coordinates [32, 33]. Another approach is the use of Park transformation with a synchronizing circuit (d-q coordinate system) to conceive control algorithms based on the dq reference frame. A comparison involving all of these algorithms for active power filters was introduced in [33].

A different methodology from the aforementioned corresponds to the active and non-active currents, which does not present any kind of coordinate translation. It derives from Fryze active current concept and presents a very simple formulation as introduced in [34]. Essentially, this algorithm determines the minimum (active) current component that transports the same energy of a generic three-phase load current. Due to its simplicity, we choose the control algorithms based on the active and non-active currents as basis to exploit the performance of the active filters, considering a power grid with unbalanced voltages and nonlinear loads.

Figure 4 presents a control algorithm for constant instantaneous active power concept, whereas **Figure 5** for sinusoidal grid current concept, with the grid voltages (v_{Sa}, v_{Sb}, v_{Sc}) replaced by the control signals pll_a, pll_b, pll_c. These signals are unitary sinusoidal waveforms synchronized with the fundamental positive-sequence component of the grid voltages

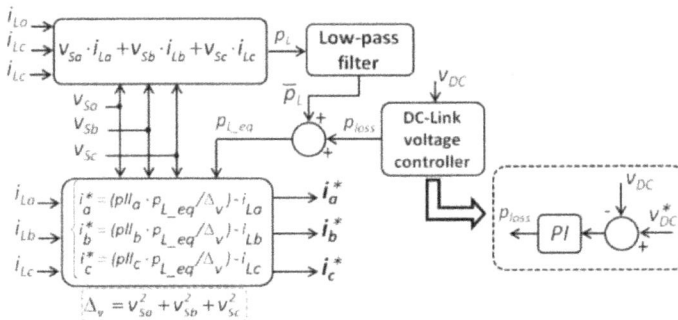

Figure 4. Control algorithm based on the Fryze active currents for constant instantaneous active power concept.

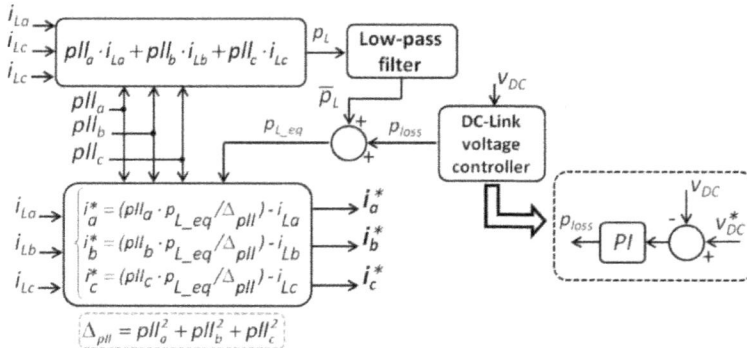

Figure 5. Control algorithm based on the Fryze active currents for sinusoidal grid current concept.

(v_{Sa}, v_{Sb}, v_{Sc}), and they were obtained through a PLL circuit [35–38]. It is important to highlight that when sinusoidal grid currents are required, considering unbalanced or distorted grid voltages, a circuit capable to extract the fundamental positive-sequence component of the grid voltages must be added to the control algorithm of the active filter, independently of the chosen methodology.

Based on both control algorithms, one can see that the control signal $\mathbf{p_L}$ presents different meanings. Indeed, for constant instantaneous power concept (**Figure 4**), $\mathbf{p_L}$ derives from the active power of the grid, whereas for sinusoidal current concept (**Figure 5**), $\mathbf{p_L}$ derives from the active power involving the fundamental positive-sequence component of the load currents only.

This issue can be better understood through the illustrated results in **Figures 6** and **7**. According to the simulation results in **Figure 6**, for providing constant active power, the compensated grid current still presents some harmonic components from the load current. It is important to comment that, according to the definitions proposed by Fryze, p_{grid} corresponds to the active power, whereas all the other components represent the non-active power. In this case study, there is only active power due to applied control algorithm.

On the other hand, as shown in **Figure 7**, the compensated current is sinusoidal even with a distorted grid voltage. Moreover, once the average component of q_{grid} is equal to zero, it is possible to affirm that the compensated current is in phase with the fundamental positive-sequence component of the grid voltage. A negative aspect of this concept is the presence of oscillating components at p_{grid} and q_{grid}, which may compromise the performance of other equipment and devices connected to this power grid, where the active power corresponds to the average component of p_{grid}, with the remaining components representing the non-active power.

Particularly, for minimizing the involved costs of the active filter, one can consider selective harmonic filtering as a feasible possibility. In this case, the compensation of a few harmonic components, especially the lower harmonic orders (third and fifth harmonics, for instance), may result in the compensated grid current with a total harmonic distortion (**THD**) lower than

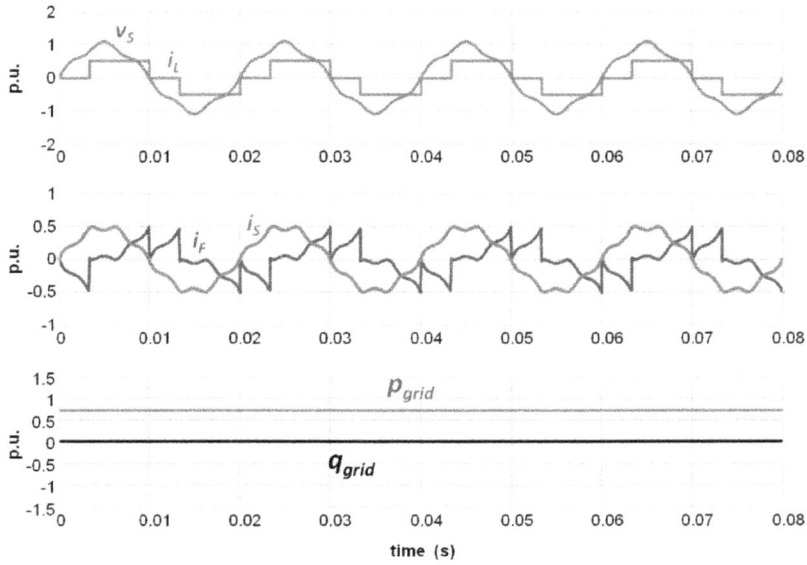

Figure 6. Simulation results with the control algorithm based on the constant active power concept.

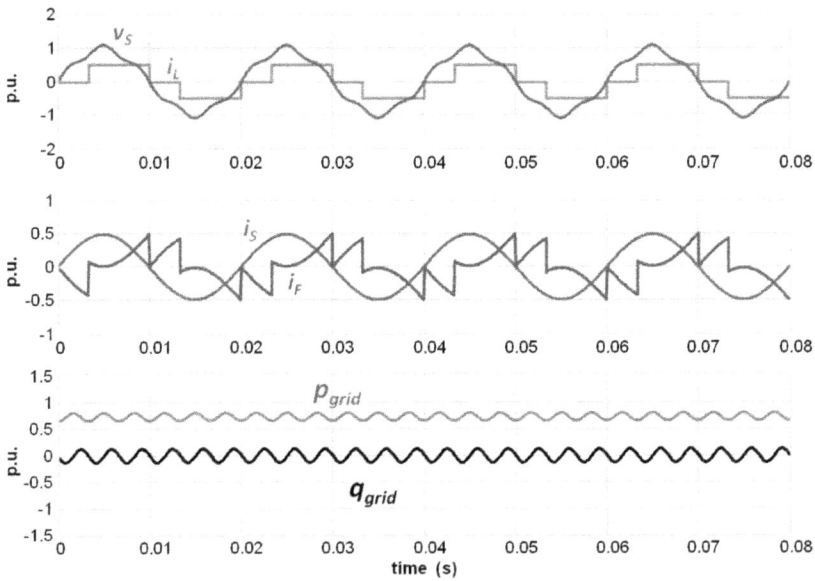

Figure 7. Simulation results with the control algorithm based on the sinusoidal grid current concept.

5%, which is acceptable for most of power quality norms and recommendations. Other possibility is to replace the compensation of a specific harmonic component by a harmonic symmetrical component, in case of unbalance load currents, as proposed in [39].

2.2. Control algorithms for series active filter

As depicted in **Figure 8**, the main control algorithms to the series active filter comprehend a PLL circuit, an algorithm to extract the fundamental positive-sequence component of the grid voltages, the DC-link voltage controller, and a damping algorithm. With these control algorithms, the series active filter is able to provide full compensation of harmonics and unbalanced components; and moreover, it is also capable to improve the power grid stability through the damping controller. In sequence, the algorithm for determining the positive-sequence component of the grid voltages and the damping algorithm are exploited.

According to the block diagrams illustrated in **Figure 9**, one can see a similar methodology for determining the control signals, $v_{S1 + \_a}$, $v_{S1 + \_b}$, $v_{S1 + \_c}$, when compared with the one applied for determining the reference currents of the shunt active filter, based on the sinusoidal grid current concept.

A preliminary result of the series active filter is illustrated in **Figure 10**. With the introduced control algorithms shown in **Figure 8**, the amplitude of the compensated grid voltage is slightly decreased. It occurs due to the amount of energy necessary for keeping the DC-link voltage regulated, which is directly related to the power losses of the VSI and the small passive filters as well.

As alternative to mitigate this problem, one can consider the addition of an algorithm to obtain a controlled voltage in quadrature with the control signals v_{S1+_a}, v_{S1+_b}, v_{S1+_c}. It is important to comment that these added voltages do not produce active power with the grid currents, and, consequently, they do not interfere on the flow of energy between the active filter with the power grid. A block diagram of this algorithm is shown in **Figure 11**, where the amplitude

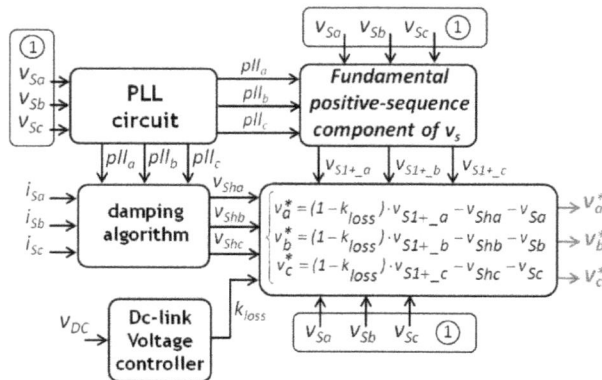

Figure 8. Block diagrams of the control algorithms to the series active filter, with damping controller and DC-link voltage controller.

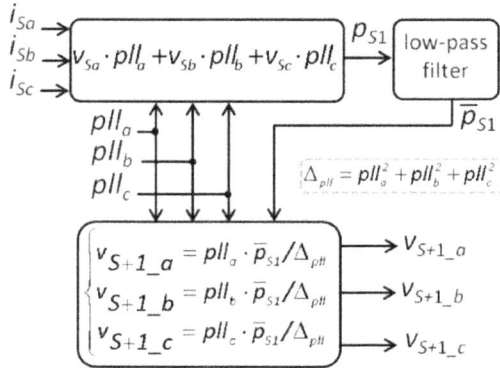

Figure 9. Block diagrams of the algorithm to determine the fundamental positive-sequence component of the grid voltage.

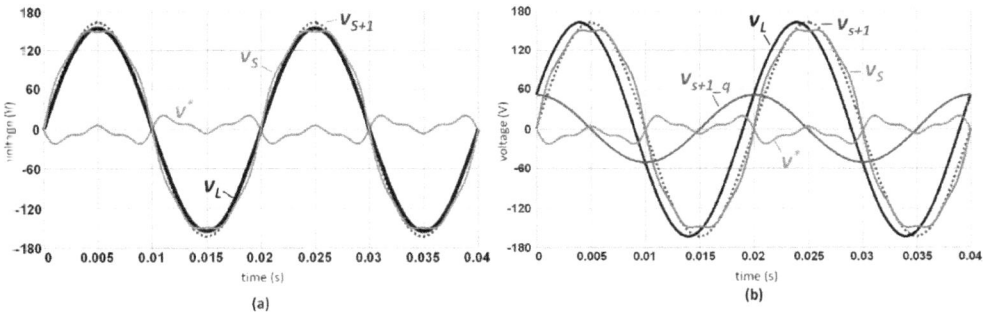

Figure 10. Preliminary results of the series active filter (a) with the control algorithms introduced in **Figure 8** and (b) adding an algorithm for compensating the drop voltage derived from the DC-link voltage controller.

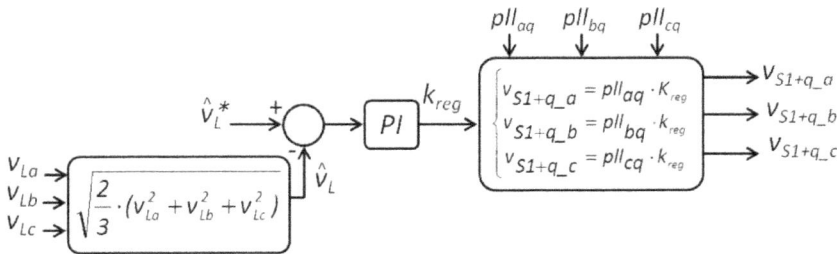

Figure 11. Block diagrams of the algorithm for determining controlled voltages in quadrature with the fundamental positive-sequence component of the grid voltages.

reference of the load voltages is compared with their aggregated value, being the amplitude of the controlled voltages determined by this algorithm. Furthermore, the control signals pll_{aq}, pll_{bq}, pll_{cq} are determined through the PLL circuit, which are unitary sinusoidal waveforms leading the control signals pll_a, pll_b, pll_c by 90°.

In case of adding tuned passive filters together with the series active filter, some constraints must be taken into account. In this topology, the passive filters provide a low impedance path to some of the harmonic components of the load currents, improving the performance of the series active filter. On the other hand, instability problems due to resonance phenomena involving the passive filters with the grid impedance may occur. An alternative to overcome this problem is to add the damping algorithm [29]. Through this algorithm, the series active filter produces a controlled voltage that behaves as a resistance to the harmonic currents that should be drawn by the passive filters.

Based on the block diagrams illustrated in **Figure 12**, the damping voltages (v_{Sha}, v_{Shb}, v_{Shc}) results from the direct product involving the non-active components of the grid currents (i_{Sha}, i_{Shb}, i_{Shc}) with the controlled signal R_h that can be understood as a controlled resistance to the non-active currents. Nevertheless, note that R_h must be designed for providing a controlled resistance to the non-active currents only. Otherwise it may compromise the flow of the active component of the grid current.

In sequence, we provide simulation results from a test case of the series active filter combined with shunt passive filters, as shown in **Figure 13**. The nonlinear load corresponds to the six-pulse thyristor bridge rectifier and the passive filters comprehend two selective passive filters at fifth and seventh harmonics, plus a passive filter for high-order harmonics. In this test case, while the active filter was turned OFF, there was a resonance among the passive filters with the grid impedance with some undesirable effects as, for example, distorted grid voltages (**Figure 13a**) When the active was turned ON, the resonance was damped in a time period lower than one cycle period (**Figure 13b**), with the active filter providing a controlled resistance to the non-active components of the grid current and, as a consequence, the active and passive filters presented a better performance as illustrated in **Figure 13c** and **d**, respectively. In this test case, at steady state, the THD of the grid currents decreased from 35% to less than 5%, which is acceptable by most of recommendations and norms related to power quality indexes.

2.3. Control algorithms for unified power quality conditioner

Essentially, the UPQC control algorithms combine those from the series and the shunt active filters with simplifications. Indeed, as illustrated in **Figure 14**, the UPQC control algorithms

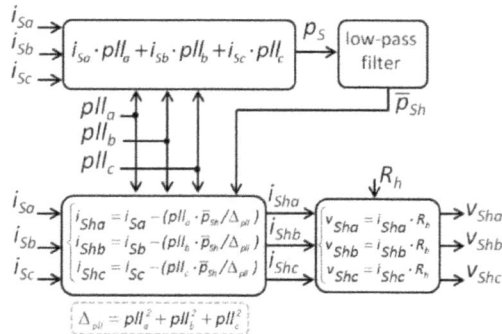

Figure 12. Block diagrams of the damping algorithm.

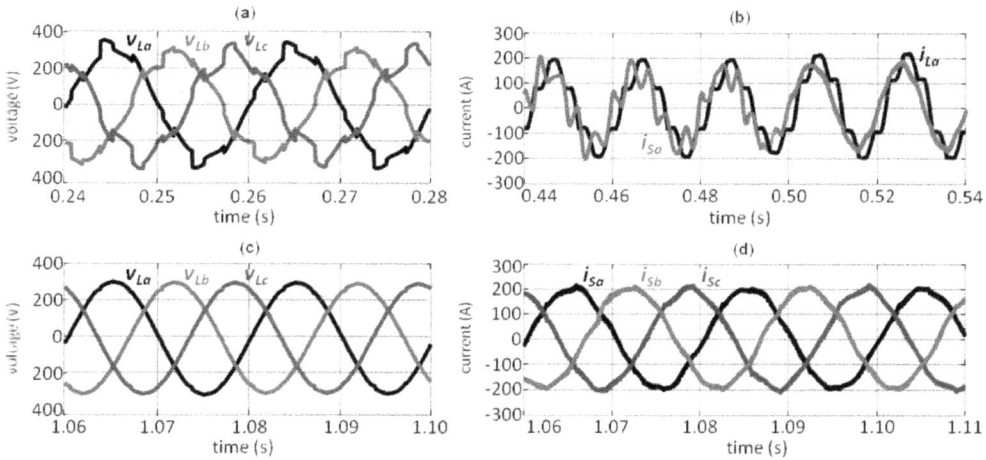

Figure 13. Simulation results of the series active filter combined with shunt passive filters; (a) load voltages with the active filter turned OFF, (b) grid and load currents at the transient when the active filter is turned ON, (c) load voltages with the active filter turned on under steady state, and (d) grid currents with the active filter turned on under steady state.

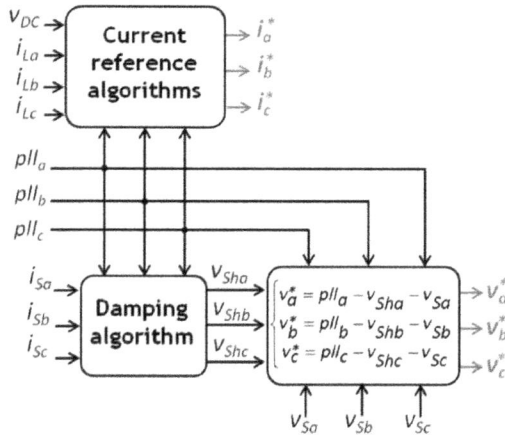

Figure 14. Block diagrams of the unified power quality conditioner.

comprehend the control algorithms of the shunt active filter, with a PLL circuit and the damping algorithm. The reference voltages are determined from a combination involving the grid voltages and the output signals of the damping algorithm and the PLL circuit.

Note that the algorithm to determine the fundamental positive sequence of the grid voltages was removed, with their outputs replaced by the PLL output signals. Indeed, if the measured system voltage is normalized such that an unity amplitude represents its nominal value, this normalized voltage signal can be directly compared with the PLL output to achieve the compensating voltage references. In this case, the difference between the PLL outputs and the normalized voltages includes also sags or swells, as well as imbalances and distortions, which may be affecting the grid voltages.

Basically, to cover the power losses of the UPQC converters and the compensation of voltage sag or voltage swell, the shunt active filter produces a controlled current to keep the DC-link voltage regulated, with the amplitude of the grid currents being dynamically modified according to the UPQC power losses and the short duration voltage variations (SDVVs) compensated the series active filter as well.

Simulation results exploiting the UPQC compensation capabilities are shown in **Figures 15** and **16**. The nonlinear load corresponds to the 12-pulse thyristor bridge rectifier, and an unbalanced load was connected and removed from the power grid. One can see the capability of the shunt active filter compensating the harmonic and unbalance components of the load currents, with the compensated grid currents with low harmonic distortion (THD lower than 3%) and balanced. There is a dynamics at the amplitude of the grid currents due to the low-pass filter and the DC-link voltage controller as well. Based on the acquired results, it has taken more than 100 ms to the grid currents to reach their novel steady-state condition when a transient at the load current has occurred.

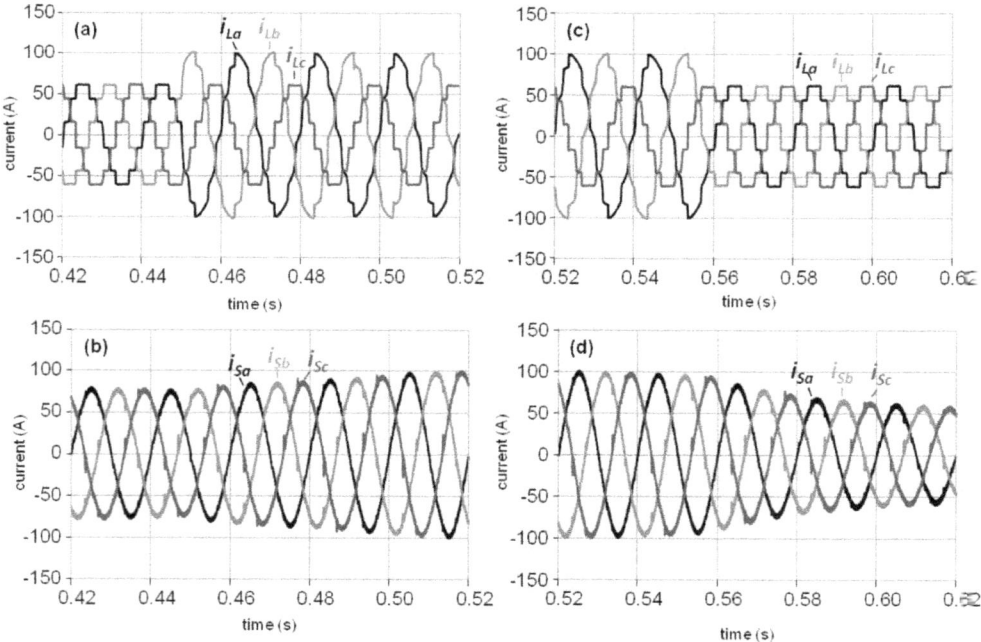

Figure 15. Simulation results of the UPQC shunt converter: (a) distorted load currents at the time transient when the unbalanced load was connected; (b) compensated grid currents at the transient when the unbalance load was connected; (c) distorted load currents at the transient when the unbalanced load was removed; and (d) compensated grid-currents at the transient when the unbalanced load was removed.

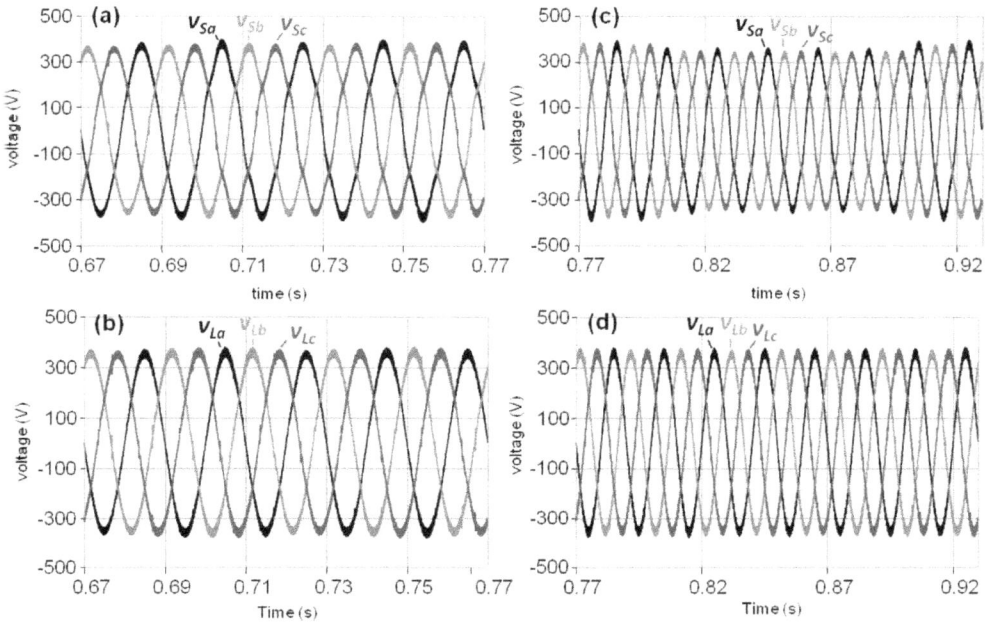

Figure 16. Simulation results of the UPQC series converter: (a) distorted grid voltages at the time transient when they become unbalanced; (b) compensated load voltages at the transient when the distorted grid voltages become unbalanced; (c) distorted and unbalanced grid voltages at the transient when a voltage sag occurs; (d) compensated load voltages at the transient when a voltage sag occurs.

Figure 16 illustrates the performance of series active filter compensating harmonic and unbalanced components at the grid voltage and a voltage sag occurrence. It can be noted a faster dynamic response of the series active filter, in comparison with the shunt active filter, once the series active filter is not affected by the DC-link voltage dynamics, enabling a quasi-instantaneous capability for transient compensation as shown in **Figure 16b** and **d**. In this section, we could verify the capability of the active filters for compensating most of the power quality problems. Nevertheless, there is another feature of them considered to be as interface for renewable energy sources, particularly, to the photovoltaic panels and wind systems as extremely diffused in the literature. This issue is exploited in the next section.

3. Integrating active power filters with renewable energy sources

Researches on high-performance power electronic converters combined with renewable energy sources (RENs) capable to extract more energy at a lower cost leads this technology to become technically and economically feasible to meet all the global energy needs. Encompassed by this course of events, there is a novel tendency for replacing the conventional centralized generation systems, with long transmission lines, to the distributed generation (DG) systems. In this novel concept on DG systems, renewable energy sources and storage systems are combined with the existent conventional sources to supply stand-alone or grid-connected loads. Moreover, it provides a better way of using onsite energy resources, minimizing transmission and distribution costs, which is crucial to reduce obstacles for rural or remote areas electrification and to encourage sustainable business development.

In this scenario, which comprehends the real modern power grids, active filters play a key role as an interface for connecting the REN to the power grid. For example, consider the shunt active filter illustrated in **Figure 1** with photovoltaic panels and a DC-DC converter connected at the DC-link voltage, as indicated in **Figure 17**, and the shunt active filter presents an additional feature of controlling the produced energy of the photovoltaic panels to the power grid. Usually, there is a boost converter between the photovoltaic panels and the DC-link voltage (v_{DC}), once the terminal voltage on these panels is much lower than v_{DC}.

For extracting the maximum energy of these panels, the maximum power point tracking (MPPT) algorithm controls the duty cycle of the boost converter. Through the combined operation between the MPPT algorithm and the DC-link voltage controller, it is possible to control the exchange of energy from the PV to the power grid [40]. Consider the output signal of the DC-link voltage controller, labeled in **Figure 4** as P_{loss}, to understand this dynamics. In this scenario, once the produced current by the PV arrays exceeds the active power losses of the converters, P_{loss} becomes naturally negative. In this power balance, the duty cycle control of the MPPT algorithm increases while the derivative of the PV active power is positive, with the control signal P_{loss} becomes more negative to keep the DC-link voltage regulated at its rated value. This interactive loop stops when the derivative of the PV active power is equal to zero, which means that the optimal set point (MPPT) was reached. To avoid loss of controllability, it is recommended to include an enable condition to update the duty cycle output of the MPPT

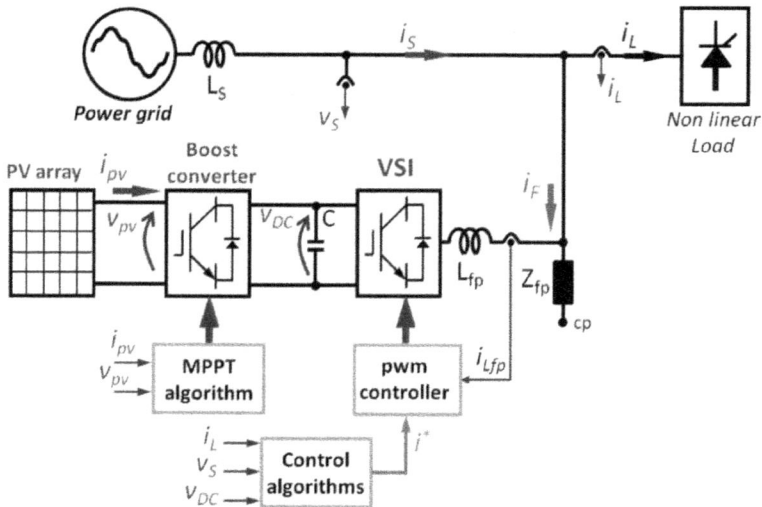

Figure 17. Shunt active filter as an interface for connecting photovoltaic panels to the power grid.

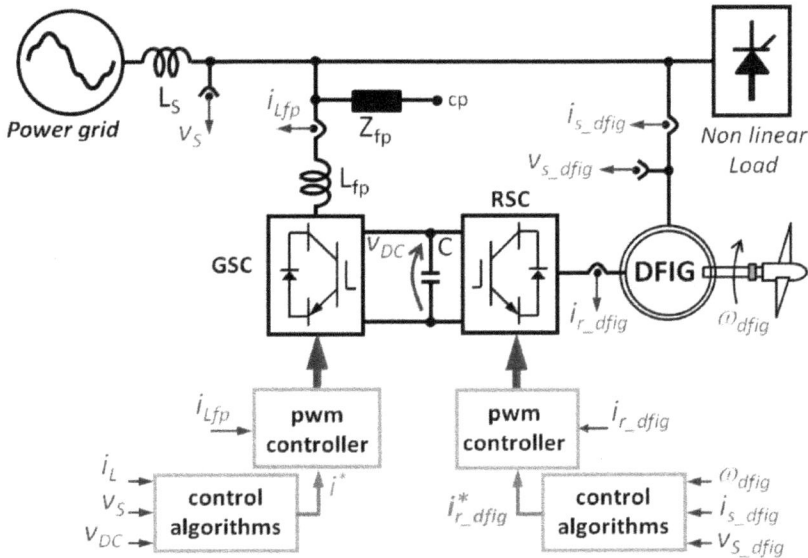

Figure 18. Simplified scheme of back-to-back converters with doubly fed induction generator (DFIG) wind turbine.

algorithm only when P_{loss} reaches its steady-state condition. Another alternative is to consider the DC-link voltage controller with a faster dynamics in comparison to the MPPT algorithm.

Another possibility is integrating active filters with Doubly-Fed Induction Generator (DFIG) wind turbine as illustrated in **Figure 18**, with one converter connected to the DFIG (RSC−rotor side converter) and the other one presents shunt connection to the power grid (GSC−grid side converter). It is notorious that RSC controls the flow of energy from DFIG to the DC-link, whereas GSC transfers the stored energy on DC-link to the power grid.

In this configuration, the MPPT algorithm is included on RSC control algorithms. Its input is the mechanical speed (ω_{dfig}) with the corresponding produced active power as the output. Moreover, the objective of RSC is to control the reactive power in the stator and the total active power of the DFIG (rotor and stator active powers), controlling the energy flow between the generator and the DC-link voltage. On the other hand, GSC produces controlled currents in counter phase with the grid voltages due to the DC-link voltage controller.

Backing to the MPPT algorithm, a possible algorithm corresponds in incrementing the reference current component related to the rotor active power ($i^*_{r\_dfig}$) while the derivative of the active power of the generator is positive. In the literature, there are several proposals of MPPT algorithms for wind energy systems as described in [41].

Nevertheless, there are some bottlenecks for connecting RENs to the power grid. One of them is their intermittent behavior, resulting in voltage- and frequency-deviations, which means an oscillating energy flow. This feature is usual in weak systems, where low inertia dispatchable power source and highly variable RENs are expected. This problem can be mitigated, under certain limits, through the shunt active filters with an energy storage element, capable to confine the oscillating energy between the active filter and the load [31].

Other issue is the load power sharing between different power converters in a decentralized microgrid. An alternative to overcome this problem is extending the droop controller concept to the shunt active filters connected in the same power grid. In this case, the active filters modify their output impedance through the virtual impedance method [42]. Basically, once these active filters share the same grid voltage, they are conditioned to produce controlled currents such that their output impedance is modified according to the capabilities of sharing the active- and reactive powers of the load. This issue is one of the most exploited ones by researchers to make the implementation of decentralized microgrids reliable.

4. Conclusions

Through this chapter, one can see the active filters capability for improving the power quality indexes due to their capability of producing, in almost real-time, controlled currents and voltages as verified through simulation results of the shunt active filter, series active filter, and the unified power quality conditioner compensating different power quality problems. Moreover, they play a key role integrating RENs to the power grid in a new concept of distributed generation (DG) systems, conditioning them to produce their maximum available energy through MPPT algorithms. Finally, the new concepts of virtual impedance algorithms allow connecting several active powers in the same power grid running autonomously in decentralized microgrids, in a similar way as the generation systems sharing the load power.

Author details

Luís Fernando Corrêa Monteiro

Address all correspondence to: lmonteiro@uerj.br

Department of Electronics and Communications, Rio de Janeiro State University, Rio de Janeiro, Brazil

References

[1] Li H, Liu C, Li G, Annakkage U. Screening technique for identifying the risk of subsynchronous resonance. IET Generation, Transmission & Distribution. 2016;10:1589-1596. DOI: 10.1049/iet-gtd.2015.0764

[2] Sahraei-Ardakani M, Hedman KW. Computationally efficient adjustment of FACTS set points in DC optimal power flow with shift factor structure. IEEE Transactions on Power Systems. 2017;32:1733-1740. DOI: 10.1109/TPWRS.2016.2591503

[3] Ghahremani E, Kamwa I. Analysing the effects of different types of FACTS devices on the steady-state performance of the Hydro-Québec network. IET Generation, Transmission & Distribution. 2014;**8**:233-249. DOI: 10.1049/iet-gtd.2013.0316

[4] Gonzalez JM, Canizares CA, Ramirez JM. Stability modelling and comparative study of series vectorial compensators. IEEE Transactions on Power Delivery. 2010;**25**:1093-1103. DOI: 10.1109/TPWRD.2009.2034905

[5] Wang L, Lam CS, Wong MC. Modelling and parameter design of thyristor-controlled LC-coupled hybrid active power filter (TCLC-HAPF) for unbalanced compensation. IEEE Transactions on Industrial Electronics. 2017;**64**:1827-1840. DOI: 10.1109/TIE. 2016.2625239

[6] Brandão DI, Guillardi H, Morales-Paredes HK, Marafão FP, Pomilio JA. Optimized compensation of unwanted current terms by AC power converters under generic voltage conditions. IEEE Transactions on Industrial Electronics. 2016;**63**:7743-7753. DOI: 10.1109/TIE.2016.2594226

[7] Marafão FP, Brandão DI, Costabeber A, Morales-Paredes HK. Multi-task control strategy for grid-tied inverters based on conservative power theory. IET Renewable Power Generation. 2015;**9**:154-165. DOI: 10.1049/iet-rpg.2014.0065

[8] Khederzadeh M, Sadeghi M. Virtual active power filter: A notable feature for hybrid ac/dc microgrids. IET Generation, Transmission & Distribution. 2016;**10**:3539-3546. DOI: 10.1049/iet-gtd.2016.0217

[9] Chilipi RR, Al Sayari N, Beig AR, Al Hosani K. A multitasking control algorithm for grid-connected inverters in distributed generation applications using adaptive noise cancellation filters. IEEE Transactions on Energy Conversion. 2016;**31**:714-727. DOI: 10.1109/TEC.2015.2510662

[10] Zheng C, Zhou L, Yu X, Li B, Liu J. Online phase margin compensation strategy for a grid-tied inverter to improve its robustness to grid impedance variation. IET Power Electronics. 2016;**9**:611-620. DOI: 10.1049/iet-pel.2015.0196

[11] Azzouz MA, El-Saadany EF. Multivariable grid admittance identification for impedance stabilization of active distribution networks. IEEE Transactions on Smart Grid. 2017;**8**:1116-1128. DOI: 10.1109/TSG.2015.2476758

[12] Zhou L et al. Robust two degrees-of-freedom single-current control strategy for LCL-type grid-connected DG system under grid-frequency fluctuation and grid-impedance variation. IET Power Electronics. 2016;**9**:2682-2691. DOI: 10.1049/iet-pel.2016.0120

[13] Chen X, Zhang Y, Wang S, Chen J, Gong C. Impedance-phased dynamic control method for grid-connected inverters in a weak grid. IEEE Transactions on Power Electronics. 2017;**32**:274-283. DOI: 10.1109/TPEL.2016.2533563

[14] Tani A, Camara MB, Dakyo B. Energy management in the decentralized generation systems based on renewable energy—Ultracapacitors and battery to compensate the wind/load power fluctuations. IEEE Transactions on Industry Applications. 2015;**51**:1817-1827. DOI: 10.1109/TIA.2014.2354737

[15] Chen L et al. Coordinated control of SFCL and SMES for transient performance improvement of microgrid with multiple DG units. Canadian Journal of Electrical and Computer Engineering. 2016;**39**:158-167. DOI: 10.1109/CJECE.2016.2520496

[16] Silva-Saravia H, Pulgar-Painemal H, Mauricio JM. Flywheel energy storage model, control and location for improving stability: The Chilean case. IEEE Transactions on Power Systems. 2017;**32**:3111-3119. DOI: 10.1109/TPWRS.2016.2624290

[17] Galeshi S, Iman-Eini H. Dynamic voltage restorer employing multilevel cascaded H-bridge inverter. IET Power Electronics. 2016;**9**:2196-2204. DOI: 10.1049/iet-pel.2015.0335

[18] Komurcugil H, Biricik S. Time-varying and constant switching frequency-based sliding-mode control methods for transformerless DVR employing half-bridge VSI. IEEE Transactions on Industrial Electronics. 2017;**64**:2570-2579. DOI: 10.1109/TIE.2016.2636806

[19] Oliveira SVG, Barbi I. A three-phase step-up DC–DC converter with a three-phase high-frequency transformer for DC renewable power source applications. IEEE Transactions on Industrial Electronics. 2011;**58**:3567-3580. DOI: 10.1109/TIE.2010.2084971

[20] Tolbert LM, Peng FZ, Habetler TG. A multilevel converter-based universal power conditioner. IEEE Transactions on Industrial Applications. 2000;**36**:596-603. DOI: 10.1109/28 833778

[21] Lee WC, Lee DM, Lee TK. New control scheme for a unified power-quality compensator-Q with minimum active power injection. IEEE Transactions on Power Delivery. 2010;**25**:1068-1076. DOI: 10.1109/TPWRD.2009.2031556

[22] Ambati BB, Khadkikar V. Optimal sizing of UPQC considering VA loading and maximum utilization of power-electronic converters. IEEE Transactions on Power Delivery. 2014;**29**:1490-1498. DOI: 10.1109/TPWRD.2013.2295857

[23] Akagi H, Nabae A, Atoh S. Control strategy of active power filters using multiple voltage-source PWM converters. IEEE Transactions on Industry Applications. 1986; **IA-22**:460-465. DOI: 10.1109/TIA.1986.4504743

[24] Peng FZ, Akagi H, Nabae A. A study of active power filters using quad-series voltage-source PWM converters for harmonic compensation. IEEE Transactions on Power Electronics. 1990;**5**:9-15. DOI: 10.1109/63.45994

[25] Tanaka T, Akagi H. A new method of harmonic power detection based on the instantaneous active power in three-phase circuits. IEEE Transactions on Power Delivery. 1995;**10**:1737-1742. DOI: 10.1109/61.473386

[26] Cheng PT, Bhattacharya S, Divan D. Operations of the dominant harmonic active filter (DHAF) under realistic utility conditions. IEEE Transactions on Industry Applications. 2001;**37**:1037-1044. DOI: 10.1109/28.936394

[27] Pigazo A, Moreno VM, Estebanez EJ. A recursive park transformation to improve the performance of synchronous reference frame controllers in shunt active power filters. IEEE Transactions on Power Electronics. 2009;**24**:2065-2075. DOI: 10.1109/TPEL.2009 2025335

[28] Newman MJ, Zmood DN, Holmes DG. Stationary frame harmonic reference generation for active filter systems. IEEE Transactions on Industry Applications. 2002;**38**:1591-1599. DOI: 10.1109/TIA.2002.804739

[29] Monteiro LFC, Aredes M, Pinto JG, Exposto BF, Afonso JL. Control algorithms based on the active and non-active currents for a UPQC without series transformers. IET Power Electronics. 2016;**9**:1985-1994. DOI: 10.1049/iet-pel.2015.0642

[30] Moreno VM, Pigazo A. Modified FBD method in active power filters to minimize the line current harmonics. IEEE Transactions on Power Delivery. 2007;**22**:735-736. DOI: 2007.10. 1109/TPWRD.2006.886769

[31] Monteiro LFC, Afonso JL, Pinto JG, Watanabe EH, Aredes M, Akagi H. Compensation algorithms based on the p-q and CPC theories for switching compensators in micro-grids. In: Proceedings of the Brazilian Power Electronics Conference (COBEP'09), 2009. Mato Grosso, Brazil, pp. 32–40. DOI: 10.1109/COBEP.2009.5347593

[32] Herrera RS, Salmeron P. Instantaneous reactive power theory: A comparative evaluation of different formulations. IEEE Transactions on Power Delivery. 2007;**22**:595-604. DOI: 10.1109/TPWRD.2006.881468

[33] Herrera RS, Salmeron P, Kim H. Instantaneous reactive power theory applied to active power filter compensation: Different approaches, assessment, and experimental results. IEEE Transactions on Industrial Electronics. 2008;**55**:184-196. DOI: 10.1109/TIE.2007.905959

[34] Xu Y, Tolbert LM, Chiasson JN, Campbell JB, Peng FZ. A generalised instantaneous non-active power theory for STATCOM. IET Electric Power Applications. 2007;**1**:853-861. DOI: 10.1049/iet-epa:20060290

[35] Konstantopoulos GC, Zhong QC, Ming WL. PLL-less nonlinear current-limiting controller for single-phase grid-tied inverters: Design, stability analysis, and operation under grid faults. IEEE Transactions on Industrial Electronics. 2016;**63**:5582-5591. DOI: 10.1109/ TIE.2016.2564340

[36] Hadjidemetriou L, Kyriakides E, Blaabjerg F. An adaptive tuning mechanism for phase-locked loop algorithms for faster time performance of interconnected renewable energy sources. IEEE Transactions on Industry Applications. 2015;**51**:1792-1804. DOI: 10.1109/ TIA.2014.2345880

[37] Carneiro H, Monteiro LFC, Afonso JL. Comparisons between synchronizing circuits to control algorithms for single-phase active converters. In: Proceedings of the IEEE Conference on Industrial Electronics (IECON'09); 3229–3234, 2009. Porto, Portugal. DOI: 10. 1109/IECON.2009.5415214

[38] Karimi-Ghartemani M et al. A new phase-locked loop system for three-phase applications. IEEE Transactions on Power Electronics. 2013;**28**:1208-1218. DOI: 10.1109/TPEL. 2012.2207967

[39] Freitas CM, Monteiro LFC, Watanabe EH. A novel current harmonic compensation based on resonant controllers for a selective active filter. In: Proceedings of the IEEE Conference

on Industrial Electronics (IECON'16), 2016, Florence, Italy, pp. 3666-3671. DOI: 10.1109/IECON.2016.7793519

[40] Perera B, Ciufo P, Perera S. Advanced point of common coupling voltage controllers for grid-connected solar photovoltaic (PV) systems. Renewable Energy. 2016;**86**:1037-1044. DOI: 10.1016/j.renene.2015.09.028

[41] Abdullah MA et al. A review of maximum power point tracking algorithms for wind energy systems. Renewable and Sustainable Energy Reviews. 2012;**16**:3220-3227. DOI: 10.1016/j.rser.2012.02.016

[42] Zhang Y, Yu M, Liu F, Kang Y. Instantaneous current-sharing control strategy for parallel operation of UPS modules using virtual impedance. IEEE Transactions on Power Electronics. 2013;**28**:432-440. DOI: 10.1109/TPEL.2012.2200108

www.ingramcontent.com/pod-product-compliance
Lightning Source LLC
Chambersburg PA
CBHW081236190326
41458CB00016B/5808